About the author

Some years ago, I gave up secure employment to concentrate fully on writing. I don't believe anyone can be really happy unless doing the type of work they love best. The journey has been financially difficult but spiritually rewarding. Having spent too much time seeking the right genre, I reverted to my first love, philosophy. This work represents the culmination of that long quest. I call it Primate Philosophy, a new thought discipline. It details how humans are not in control of their actions nor destiny, but are governed by the laws of nature that determine the course of their lives. They are likewise controlled by the legacy of their evolution as primates. There can be no escaping that primate legacy nor its heritage.

John J Mulhall

MAN AND MOTHER NATURE

John J Mulhall

MAN AND MOTHER NATURE

Vanguard Press

VANGUARD PAPERBACK

© Copyright 2021
John J Mulhall

The right of John J Mulhall to be identified as author of this work has been asserted by him in accordance with the Copyright, Designs and Patents Act 1988.

All Rights Reserved

No reproduction, copy or transmission of this publication may be made without written permission.
No paragraph of this publication may be reproduced, copied or transmitted save with the written permission of the publisher, or in accordance with the provisions of the Copyright Act 1956 (as amended).

Any person who commits any unauthorised act in relation to this publication may be liable to criminal prosecution and civil claims for damages.

A CIP catalogue record for this title is available from the British Library.

ISBN 978 1 784658 69 4

Vanguard Press is an imprint of
Pegasus Elliot MacKenzie Publishers Ltd.
www.pegasuspublishers.com

First Published in 2021

Vanguard Press
Sheraton House Castle Park
Cambridge England

Printed & Bound in Great Britain

Dedication

For Mick and Ellen

INTRODUCTION

Approximately three hundred and fifty thousand years ago, perhaps later or perhaps earlier, an evolved ape emerged timidly into a world filled with dangers. He had no natural defences to help him survive in a very dangerous world. He could not outrun the big cats, nor outfight his stronger chimpanzee cousins. Neither had evolution given him wings to escape danger by taking flight to the sky and relative safety.

Yet in a relatively short passage of evolutionary time, he has become the most dominant animal that has ever lived on the planet. He is now the fastest animal on land that has ever lived. He can fly higher than any bird that has ever flown, and is the only creature capable of leaving the planet unassisted. He can remain submerged in the oceans for months on end without having to surface for air. His eyes are now firmly set on the stars.

These remarkable achievements were made possible by his ingenious brain that has allowed him to construct machines to transport him to places that are out of bounds to his fellow creatures who share the planet with him. He has stood on the moon and sent probes to distant planets in the solar system. He is drawing up plans to colonise another planet in the solar system, and building engines to accomplish that goal.

There is no boundary to his unlimited imagination and thirst for adventure.

We know that evolved ape as Homo Sapiens, or man.

The reason man stands alone as the undisputed master of the world is due to that remarkable organ in his head called the brain. When the first human emerged from the process of evolution to begin that remarkable ascent, he had no universal book of knowledge to draw upon, nothing to tell him how the world worked. He was terrified of lightning and other natural phenomena such as ground tremors and volcanic eruptions. He saw the sun rise in the east and settle in the west, and believed the evidence of his own eyes. His world was a complete mystery, and a very dangerous place in which to live.

Now, wind the time clock forward to the present day. Three hundred and fifty thousand years, give or take, is but a small timeline when compared to the age of geological Earth, which has been evolving for four and a half billion years. In that short period of time, he has solved the mystery of lightning and knows it is caused by electrical discharges in the atmosphere. He has discovered that volcanic eruptions and earth tremors are caused by the movement of tectonic plates as they slip and collide against each other in constant motion. He has discovered that his world is not flat but round, or strictly speaking, elliptical. He has disputed the evidence of his own eyes to prove that the sun does not revolve around Earth but that the planet orbits its nearest star.

Whereas his ancestors had relied on gathering fruits in season and hunting to survive, he can now grow enough food to feed an ever-increasing population, and store it to combat future emergencies. He has discovered penicillin, the magic bullet in combatting disease. He is waging a constant war against diseases, and continues to invent drugs to ensure a better life for people in pain. He has greatly increased his longevity, and can now expect to live twice as long as his counterpart in the Middle Ages. He has the ability to clone animals, including his own kind, and genetically engineer foodstuffs to produce larger crops. He has split the atom and discovered the building blocks of life. He has more knowledge at his fingertips than at any time in the history of the world. Man stands alone as the undisputed master of his world, and as he progresses into that bright future, his brain acquires more knowledge and demands more knowledge. His brain has a spectacular and unlimited capacity to learn and to retain that knowledge.

But the story of his astonishing success has endowed him with a sense of hubris, that his continuous rise had been ordained from the beginning. His remarkable asset, the brain, sought to explain how he had gained mastery of all other creatures and came up with a mental image of supernatural beings called gods. His brain had invented religion to make sense of a strange world that made no sense otherwise. The gods had supernatural powers and could literally move mountains. The vast majority of societies that have ever existed anywhere invented supernatural beings to explain how the world came into existence, and how they were created. The human brain is a demanding master, forever seeking more knowledge and forever demanding more answers. The invention of gods would further distance man from his roots in nature.

As his brain continued to evolve, the dogma of religion was replaced by the eclectic inquiry of science. Now he sought rational explanations for natural phenomena, and for the origins of life. These enquiries further spurred the evolution of his brain. He came to believe that science could solve everything and answer every question. He had however lost something vital in his rapid rise to the top. He had lost the spiritual, that most essential facet of mankind. That aspect of human life is as essential to his wellbeing as bread or water. Man had forsaken the awe and respect that the first humans to emerge had for Earth. The ancients worshipped the planet as a mother goddess who supplied all their needs, but modern man has lost that respect and regards Earth as simply another planet orbiting another star in another galaxy not unlike countless others. That lost spirituality has been retained by isolated tribes who live as their forefathers had always lived, isolated from modernity and civilization.

This work shall examine man's role in nature, and his relationship with Earth. His ancient forefathers made offerings to the Earth Goddess in thanks for her bounty. They regarded the planet, not as an inanimate object, but as a living and bountiful matriarch who suppled everything they needed for survival. Now modern man sees the same planet as an inanimate body orbiting a star called the sun. He believes it exerts no influence on his life, and has reached a level of sophistication where he would prefer to forget his early origins and dwell only on his present condition. Modern man can be likened to a poor boy born into poverty who has made it to the summit of fame and financial success. At the pinnacle of his success, he turns his back on the impoverished mother who gave him birth to him because she is a reminder of his past life.

This work shall also set out to prove conclusively that Earth is not just another planet orbiting a star like countless others. It shall produce hard evidence that Earth influences the daily lives of every human being who lives on its hills and in its valleys, on its savannahs and in its cities. Its influence might not be apparent, but it is there nonetheless. Earth influences how people act in their daily lives, but more importantly it influences how they think. Perhaps there are other planets out there yet to be discovered containing life forms, but it remains doubtful if another one exists that gives birth to life in the way our shared planet does. Earth is a laboratory for life, creating new forms and constantly improving others under immutable laws that govern all life forms.

The ancients worshipped the planet as the Earth Goddess, another name for Mother Nature. They regarded Earth as a feminine deity, which is most appropriate and shall be retained in this work. They understood that the planet was not only their home but the personification of Mother Nature herself. That same mother had given them life and nurtured them since birth, just as their birth mother had done.

The work shall also explain how her laws have aided mankind in its rise to the pinnacle of the animal kingdom, and still guides human behaviour even if mankind has become sophisticated and distanced itself from its primeval origins. Mother Nature is not unlike a man's birth mother whose influence remains long after he has become an adult and moved away from the family home. The lessons he has learned under the guidance of a birth mother remains with him for life, although he is probably blissfully unaware of her lasting influence. Earth is his original mother, the primeval matriarch of all life forms, the original mother goddess of all life.

Ancient man understood that nature ruled his life, and he lived according to her laws. Modern man too was born under the natural laws, and continues to be governed by those laws. He is not aware of Mother Nature's influence, but it is always there to prove that he is not a creation apart. It governs all aspects of his life because he is ruled by her four immutable laws. Nature reminds him time and again that he is not a special creature, aloof from the natural world, but is part of her grand design which ensures life on the planet continues to come forth and strive for its place in the sun. Man might like to believe he has free will and can plan his own destiny, but Mother Nature still pulls the strings. She controls his actions and his thoughts.

This work shall set out to prove that Mother Nature is not a perceived deity who dwells somewhere beyond the clouds and conjured up by man, but an unseen and numinous force that not only influences our lives but also controls our moods and our actions. The evidence shall be presented to prove the case of her existence.

MOTHER EARTH AS GODDESS

Mankind once lived, in harmony on the planet and with the planet, and was part of the natural order. Man lived as other species that co-existed with him, and was subject to the same trials of life to ensure his survival. The seasons dictated how he lived and what foods he could access; and how he could survive the lean periods without starving to death. Being an omnivore gave him a distinct advantage over other animals with specialised diets because he was not limited to a single food source, and that eclectic diet was part of his success. He hunted when the game was plentiful and gathered nuts and berries when it was not. He came to regard Earth as a second mother who provided him with all the conditions necessary to sustain life. Earth gave him a home for his family to shield him from the elements. In the beginning, he took advantage of natural places of shelter like caves, before learning to build a home with the materials the planet provided such as wood and stone. He regarded his home planet as a living entity, as a most bountiful mother, and celebrated her fecundity and bounty by carving figurines in her honour. These carvings have been excavated in Europe and Asia, demonstrating that the first humans universally worshipped Earth as a second mother.

Unlike modern man who takes Earth's natural resources and gives nothing back, nor offers any thanks for the free bounty, ancient man worshipped Earth as a mother goddess and made ritual offerings to her. She was not a distant or absent goddess who lived in the sky, but a living being who cared for them every day in much the same way as their birth mother. This belief was not confined to one specific area of the planet but stretched across continents from Europe to Asia and Africa. Gaia was the name the Greeks gave to Mother Earth; Danu was the name given to her by the Celts; and in India she was named Maimata. The excavation of figurines in Turkey and elsewhere dedicated to Mother Earth has led to the belief that a matriarchal society had once existed, a peaceful and gentle society ruled by women.

This theory has gained some traction, but it can be dismissed because it contradicts human evolution. Any theory that ignores the primate origins of the human race does not stand up to close scrutiny. Although little about pre-history can be written with any degree of certainty, this speculation can be dismissed readily because the evidence of evolution contradicts it. Mankind evolved from the primate family tree and is therefore governed by the laws of that species, and by no other laws. That unbreakable primate connection demands an alpha male at the top and does not allow for female rule. There could be no alpha female under this system of rule, and a society ruled by women would, of course, require one as leader.

Mankind evolved from a primate society dominated by males, and that has been reflected in every society and culture in the history of the world, and in every religion that has ever seen the light of day. Female societies that existed in the past might make for good fantasy or erotic material, but they can have no basis in real life because man evolved from the primate family tree that was led by a male. Tribal leaders, emperors, religious leaders, and heads of state were always and universally male, just as the majority still are. There are indigenous tribes still being discovered in South America and they have one thing in common, and that is a male leader. No tribe has ever been discovered led by a woman. The reality of a female leader is a relatively recent phenomenon, and mainly in more advanced societies.

It should also be noted that no religion appoints a female at the head of its church because it contravenes the primate code. Religions tend to be traditional and do not change as quickly or as readily as secular societies. They govern their flocks as they always have, with a male leader in the top job. The furore over the appointment of female bishops recently in the Church of England was not due to anti-female bias nor conservatism, but rather that it contravened the primate system of rule that requires male leaders. The objectors were hearing a subliminal message from the past that reminded them of their primate origins, even though they believe they were created by an omnipotent being in the sky called God.

Another widespread belief is that the Earth Goddess was consort or wife to the principal god. For example, Isis in Egypt was wife to Osiris the chief god, and Juno in Rome was wife to Jupiter the chief god. These two heavenly queens have been conflated with the Earth Goddess, but a cursory glance at their images proves that this is a misconception. They are slim and graceful and beautiful goddesses befitting a Queen of Heaven as both were regarded in their respective societies. Moreover, the belief in the Earth Goddess preceded the invention of the sky gods by many millennia. The proofs can be seen in the excavated figurines that display common attributes of fecundity and bounty, such as large bellies and heavy breasts. Some figurines have been found with multiple breasts and therefore prove they are true representations of the universal Earth Goddess, or Mother Nature, and as such representative of a bountiful mother who provides sustenance for the people, rather than artistic representations. The Earth Goddess nurtured mankind, just as the mother nourishes the child at her breast.

Man's first break with Mother Nature occurred with the discovery of farming, especially the cultivation of crops. This in turn led to a complete change of lifestyle. Now that man could grow his own food, he abandoned the nomadic way of his ancestors and settled down to live a sedentary lifestyle. Evidence of man's hunter-gatherer life can still be observed in tiny pockets of nomadic societies across the planet, a reminder of that early period in human evolution. The change from roaming nomad to settled farmer required him to construct dwellings; and these evolved into towns as the population increased and eventually evolved into cities. He used fire as an ally to clear forests and open them up for cultivation, the only animal to have harnessed it as an ally. He captured wild animal like aurochs, and over generations bred them to become docile cattle as a food supply.

Cattle did not evolve naturally but were domesticated from the wild by selective breeding. Now he had a ready source of meat and had no need to hunt. He was beginning to regard nature not as a source of plenty, but as a wild and barbaric place that had to be tamed, just as he had tamed aurochs and turned them into cattle. Settled man would not only civilize himself but also Mother Nature, the planet that had given him birth and provided food for his family. The wild lands and forests outside his towns and cities, his former home, were now seen as wild and dangerous and teeming with savage beasts. The forests were not towns or cities, and therefore were uncivilized. His change in lifestyle would set him on a course that would eventually cause him to lose touch with his roots, and finally with Mother Nature herself.

Agriculture put him on a collision course with nature and her many creatures, whereas before he had lived with them side by side. His crops had to be protected from marauding animals that sought to destroy them. The farmer who sows and reaps is unable to co-exist with wild animals as the hunter-gatherer does because his livelihood depends on his crops. His crops and animals are now his prime source of his food. Neither could he live with wolves who might attack his sheep and cattle, as still happens where they live side by side with livestock. It was necessary to vilify wolves in order to kill them, and he did it with the killing efficiency and determination only humans can muster. This policy would eventually lead to the virtual elimination of wolves from Western Europe and great swathes of North America.

The change from nomadic to sedentary lifestyle brought about a new perspective in his outlook, and he saw the sun as the giver of life. It was essential to his livelihood since bad weather could ruin his harvests and threaten his survival. The sun was now regarded as the true giver of life, not planet Earth, and deified as a male god. The civilizations that grew up around the Mediterranean Sea all worshipped the sun as a male god and gave it a prominent position in their systems of belief.

The societies that relied on the sun made offerings to guarantee its benevolence for their livelihood. The sun and Earth represented a marriage between male and female, with the male as the dominant partner to reflect how the people lived. The sun fertilised Mother Earth and she became bountiful and her children benefitted. The change from the nomad to the farmer saw the sun replace Earth as the more important deity in the marriage. As towns grew into cities, the chasm between man and nature became wider yet because he believed that anywhere outside the walls was barbarous and teeming with wild beasts ready to devour him. Agriculture allowed man to control nature, and gradually his respect for the natural world began to change. The tribes who still live as hunters and gathers retain the respect for nature that modern societies have lost. An example can be found in the clash between a society that lived in harmony with nature and one that exploited her when Europeans settled North America.

Native Americans of the plains lived as hunters and gatherers, and relied on buffalo herds for most of their needs. They understood nature and did not slaughter the animals indiscriminately, but took only what they required for survival. They used the hides to make clothes and footwear, the meat for food, and the bones and sinews for tools. European settlers, the offspring of farming and cities, had no similar respect for nature and hunted the buffalo almost to extinction for their hides alone, leaving the carcases to rot where they were killed. The break with nature that began with the discovery of agriculture resulted in ignorance of the natural world, and how the elaborate system worked.

Nature is all about balance, especially that between predator and prey. The wolf and the elk can be used to illustrate the balance, and to explain how the delicate system is balanced on a knife-edge. The wolf preys on the elk for its own existence and to rear its young, as the laws of nature demand. When elk numbers fall, pack numbers also fall. The wolf might be considered a dumb animal on four legs by sophisticated man, but it instinctively knows not to kill all the elk. That could only end in starvation and death for the pack as a whole. By breaking with nature, modern man has lost sight of this most valuable lesson. Civilized man considers himself intelligent, but does not understand this delicate balance because of his split from Mother Nature. The obvious irony is that the people he regards as uncivilized, like native tribes that live as they always have, do understand the critical balance.

When European explorers first landed in North America, they spoke in wonder of its pristine waters abundant with fish, and its sky teeming with birds. Native Americans had for tens of thousands of years lived with this abundance without disturbing the balance. One animal perhaps best summed up the bounty of that land, the carrier pigeon. In the mid-nineteenth century, a young chief spoke of entering a quiet valley and then hearing loud sounds like a crashing waterfall or a herd of galloping ponies. He looked up at the sky and saw millions of carrier pigeons coming down to roost in numbers that appeared to cover the sky. They were the most numerous birds on the continent, and perhaps in the world. In less than a century, they would be hunted to extinction by the settlers who did not understand how nature works and the checks and balances she employs to regulate the system. Not a single one survives today to darken the sky and delight the present generation of Americans. European settlers, and their lack of understanding for the natural world, wiped them out.

When man interferes with the laws of nature, chaos is the usual outcome. Another example is worth citing from North America. A few decades ago, in Yellowstone National Park, a ranger noticed that it was dying. The saplings weren't growing and the river that runs through the park was clogged with weeds. The beavers had left and the grazing animals had multiplied. He proposed a radical solution: the reintroduction of the wolf to fill the gap caused by man who had persecuted the animal and driven it from its habitat. His proposal was met with howls of protests from farmers and people who lived near the park. The indoctrination that wolves were evil had conditioned the descendants of the first settlers to view these magnificent animals as mindless killers. Nevertheless, he persisted and imported Canadian timber wolves. The results were transformational.

The park recovered because the balance of nature was restored. Animals that had been grazing on the saplings now were kept on the move by the wolves, thus allowing the young trees to mature. The herds became healthier since the wolf weeded out the sick thereby fulfilling its role in nature. The beaver returned and the river regained its former health. The apex predator had restored the balance that man had removed in his ignorance of the natural world. The wolf was not a villain as it had been portrayed, but the essential piece in the interconnected jigsaw of life. By removing the top predator, man had destroyed the essential balance that permitted diverse species to survive in the same habitat.

Native Americans had lived with the wolf for countless generations, without hunting it to extinction nor driving it from its habitat. European immigrants to the continent wanted to settle down and farm. They regarded the wolf as a threat to their livestock, but also as part of the wild wilderness that must be tamed. They were bringing civilization to the continent, without realising the havoc they were creating.

They had lost touch with nature and their origins. They did not understand the vital role played by the wolf in the ecology of the system that Mother Nature had bequeathed them. The wolf had evolved specifically to fulfil that role, and to improve both prey and predator.

Dominance always produces hubris, and by becoming the top predator on the planet man has forgotten his roots. He no longer lives in harmony with the natural world, but instead exploits it and its life forms. He takes and takes, and gives little back in return, Not only is his attitude unsustainable, it is also ignorant. It shows no respect for the planet that gave him birth, his original mother. If that is the price of civilization, perhaps man should take a lesson from his ancestors.

THE FOUR NATURAL LAWS

Earth is an orderly environment with laws governing its inhabitants and their behaviour. Species live within their own common set of laws, and are punished if the laws are transgressed. For example, a lion venturing into the territory of another lion will be attacked. Both parties are aware of the laws and know the price of transgression. Where laws are seen to exist, it must follow that a lawmaker has set them down. Mother Nature is that lawmaker, and she has established the rules under which each species lives. Just as the mother sets down the rules in her household for her children to follow, Mother Nature has set down her laws because she gave birth to all life forms on the planet.

There are four basic laws that ensure life continues on planet Earth, and mankind too obeys them since they have enabled the survival of Homo Sapiens when rival human species have become extinct. It must be assumed that they were rendered extinct because they failed to obey the laws of nature. Survival is success whereas extinction is failure, and that statement is seen to be both factual and relevant. The four inviolable natural laws are universally obeyed since they guarantee success in the battle for survival and reproduction.

The four laws of Mother Nature are:

Law I. Survive to Adulthood.
Law 2. Compete for Territory.
Law 3. Endow Offspring with Successful Genes.
Law 4: Permit No Rival Gain an Advantage.

These are the four natural laws of life that both flora and fauna adhere to and obey. The primary law is survival itself, and failure to comply with this law renders the other three redundant of course. The individual must survive to establish territory and pass down genes, the engine that propels all life on Earth. The life form that can compete best for territory is the most likely to succeed and pass down successful genes. Those unable to adequately compete risk having their genes extinguished.

Earth is a laboratory for life that births individuals within a species that are all different to each other. As such, each birth in itself is a new trial or experiment because no two life forms are identical, neither in flora or fauna. Since no adequate word exists to describe this natural system of trial and error, this work shall invent one called *neotento*. Mother Nature does not replicate or produce exact copies of life, but experiments continuously in her laboratory of life.

Under the process of *neotento,* siblings born into the same family not only are physically different but invariably have different personalities. Just as no two bodies can be the same, no two brains can be the same either. Each sibling within the same family is a new trial for life in the ongoing evolution of its particular species. Two brothers might share the same parents and similar genes, but they will be different physically and mentally. Sibling rivalry is another example of that difference, and it exists in most animal species. This rivalry is a natural desire for dominance, and is part of the evolutionary law that governs life.

The young of simian species play games that form part of the journey into adulthood. Schools have playgrounds where children can enjoy themselves between the boredom of lessons, and their enjoyment is obvious to anyone in the vicinity. Playing is part of nature's grand plan because it is a preparation for adulthood, and allows the young to establish a pecking order that adds to the cohesion of the tribe. Modern life and democratic societies might preach equality to all, but play is an essential part of the natural world and a legacy from mankind's origins that remains a vital part of childhood. Children play because they are part of the natural world, and play is a rite of passage to the world of adults.

Another and less savoury aspect of the schoolyard is the bully which is seldom out of the headlines. Each time it rears its head, the usual sociologists and child experts are wheeled out to explain why it's happens. Invariably, enough heat is expended in the discussions to warm a mansion in a snowstorm. Some blame it on bad parenting, others on socio-economic factors. Science cannot deliver answers when divorced from Mother Nature, but tends to ignore her time and again because modern man has abandoned his roots. He has become so completely detached from his origins that he doesn't even consider why bullying still persists in the twenty-first century. Modern man has reached an advanced level of civilization, but in so doing has left his understanding of the natural world in the past.

Bullying is part of the natural pecking order to establish dominance. It can never be eliminated, just as the coccyx bone can never be eliminated. It is part of the evolutionary process that transported the human race into the modern world. Children will always bully because it is part of their evolution, and the invention of computers and smart phones has made the practice easier because now the bully can act anonymously. He or she always picks on the more sensitive and vulnerable in the group, precisely because that young person is vulnerable. It's much easier to establish dominance over a sensitive child than over a confident child who has the courage to fight back.

The effects can be devastating and lead to suicide in some cases. The bully does not comprehend the damage he or she is doing to the victim, but heeding the siren call of nature to establish dominance. Yet the cause is not even contemplated or discussed because of the belief that scientific methods have all the answers. That attitude is akin to seeking the cause of potato blight by examining an orange tree in a laboratory by dissecting its roots. The proper study of mankind and its behaviour lies in its evolution, and the underlying cause of bullying is a legacy from the past where the young jockeyed for hierarchy.

Lion cubs engage in play-fighting and bullying to sort out their place in the pride, as do the young of monkeys to establish their place in the troop. By sorting out the hierarchy in a group when young, it allows every individual to find its place in that society in adult life. The school bully is usually the biggest and toughest boy or girl in the yard, which itself is a throwback to the early origins of mankind when brute strength ruled. That system still pertains in gorilla society where the strongest male rises to the top. Man's primeval ancestors passed through this phase too, but evolved from a creature of brawn to a creature of brain.

When man's ancestors evolved from a physical creature who relied on strength to a cerebral creature who relied on brain, brute strength evolved into cerebral power. In other words, brawn was replaced by brain because a clever mind offered a better chance of getting to the top than brute strength. This change too is reflected in the normal transition from childhood to modern adulthood, where bullying is no longer socially acceptable. Yet the practice persists in the workplace by some adults because it grants them a whiff of power. And another name for power is of course dominance by the bully over the bullied. The penalties for bullying by adults in the office or other workplaces can be severe, but that has not stopped it from happening because it remains part of man's heritage from the past.

Since the primary law of nature is survival, every life form on the planet obeys this command. The right to self- defence is ingrained in the individual. A porcupine uses its quills as protection and can inflict serious wounds on an attacker. A warthog attacked by a leopard will kill the cat if possible to save its own life. That inalienable right to self-defence is universal, and is universally obeyed. The first law of nature grants all creatures the right to fight for their lives when threatened. The varied array of weapons that animals have evolved enable them to obey the primary law. Huge teeth on a baboon can inflict serious damage or death on an attacker, and the horns on an antelope can also be used for defence. The thick layers of skin on rhinos and elephants is also a form of self-defence, not unlike the armour plating worn by medieval knights.

Unconsciously perhaps, man too lives by the primary law of nature. A man who kills an attacker in self-defence suffers no retribution from the laws of the state. If his life is in danger from attack and facing death, he is allowed to fight back and take the life of his attacker. The right to protect his own life ensures he is not charged with a crime. Should he kill without proving he acted in self-defence, he faces a charge of murder and life in jail. In this scenario, the law of mankind is in synch with the first law of Mother Nature.

The secondary law of Mother Nature is establishing a territory, an essential requirement for food supply and a place to birth the next generation. Territorial animals declare ownership of their patch by scent-marking it at the boundary to warn off rivals who want to take it over. A good territory is always worth fighting for because it can often mean the difference between life and death. Having secured the territory, it must be defended against all-comers or it will be lost. Man is highly territorial, and defence of that territory is also hardwired into the human psyche as can be seen by the countless wars that have raged since man's ancestors first invaded his neighbour's territory, and are still waging.

Fundamentally, all wars are for territory and always have been regardless of what the combatants on both sides claim or say. Wars of religions and wars of ideologies and wars of noble causes are merely excuses for land grabs. These excuses are used to cover the real reason for the war. After a conflict of arms ends, territory is divided up between the victors as a reward for success. The loser cedes territory and the victor gains territory, proving that territory is the name of the game.

The disputes between neighbours over a metre of ground or a right of way are often seen as trivial, until the courts have to intervene or someone is killed. A neighbour could spend a fortune in court defending a patch of land that is worth very little, and bring the dispute to the highest court in the land at enormous cost and expense in time and money. The intrinsic value of the land is not really the issue but the determined defence of his territory, and that is a primate legacy no one can ignore. Though modern, man continues to be governed by the second natural law because it's part of his origins, just as his childhood is part of his origins.

The defence of territory is hardwired in the human psyche and may be observed by the uses of obstacles such as rivers and mountains as defensive barriers. The Pyrenees form a barrier between Spain and France, and the Alps between Italy and Austria, and also other countries. The Rio Grande forms a barrier between the USA and Mexico, and the Rhine between Switzerland and Liechtenstein. These are visual barriers, and reinforced at the border or entry points by signals of ownership such as national flags. At a more local level, the home itself is sign of personal territory that will be defended by most owners if broken into by a thief intent on stealing or an intruder intent on harming the owner. Such an unwanted and illegal break-in is an invasion of personal territory that contravenes primate law and will not be tolerated. All primates are programmed by nature to defend their territory if attacked, and of course man is a primate.

The third law is reproduction, and a good territory is the best way to attract a mate. The female is attracted to male who can acquire and hold a good territory because it gives her offspring the best start in life. In the past, the human female was attracted to the male with the best territory and who could defend it from attack. She did not have to forage for food in territories owned by rivals who might have killed her offspring. By mating with the leader who held a good territory, she was ensuring the future of her offspring who would inherit the territory and immortalize her genes.

The third law still influences modern men and women subliminally because they are part of the natural order. The millionaire who can afford a luxury pad in the best part of town is widely regarded as more successful than the poor man who struggles to put a deposit on a house. To put it in evolutionary terms, he is regarded as having better genes. Fashion magazines and newspapers do not chronicle the lives of the poor after all, they write about the lives of the rich. A rich man is always more attractive to women than a poor man because his wealth proves he can give her offspring a better start in life. That is the underlying reason in choosing a mate, giving the next generation the best start in life and it applies across all species. Genes drive life in the same way an engine drives a motor car.

Young women who marry rich men old enough to be their grandfathers are usually ridiculed as mere gold-diggers and only after them for their money. But they are unconsciously obeying the third law by securing the future of their offspring. They are listening to that siren call from the past to ensure that their offspring are given the best chance in life. In the modern world, money is the equivalent of good territory and that is always worth striving for. It also goes without saying that money can also buy good territory.

The genetic imperative of the third natural law that drives life is universal and transcends both flora and fauna. The need to reproduce is as strong in the oak tree as it is in the ostrich. Dominance is a sign of good genes and the dominant produce the most offspring. This is nature's way of ensuring the best genes are passed down to the next generation. The incremental benefits as good genes are passed down from generation to generation helps the species as a whole. An advantage gained by the parent over a rival is thus bequeathed to the offspring giving it an advantage.

The laboratory of life is based on trial and error, and this too is part of nature's immutable law. Something that works is honed and refined, and passed down. Something that does not work, or is unsuccessful, is weeded out. Nature has installed a mechanism for this job and it is called the predator. If the innovation works and is successful, the recipient can evade the predator and survive, and pass on its good genes. If the innovation does not work, the recipient is eliminated by the predator. The arms race between predator and prey improves both species, as both have to react to any advantage gained by the other.

Nature's laboratory is the driving force of evolution, and can be observed in the world of business. Without perhaps realising it, modern businesses emulate the laboratory of life that has propelled evolution since the first creatures appeared on the planet. Every new product is based on trial and error, ranging from motor cars to chocolate and soft drinks.

To prove that business imitates life, a closer look should be taken at the automobile industry. If a new model of motor car is successful, the model is ramped up and production increased to meet the growing demand. Subtle changes are made to the model to meet the demands of new customers, but the same name is retained. The evolution continues with periodic face-lifts and new technology until the car is no longer distinguishable from the original model. The successful name sells the car simply because it has been a success in the past. The company advertises the success and claims the model is the best- selling car in history, even though the car has undergone many changes and bears no resemblance whatever to the original. Under the same law of evolution, if a new model is not successful and does not sell in sufficient quantities, it is dropped.

The same law applies to chocolate bars and soft drinks, and every other product. Those that sell are promoted and become part of the company's legacy, and are passed down to the next owners in the same way that evolutionary advances are passed down to the next generation. The products that do not sell in sufficient quantities, or that do not gain an advantage over the competition, are dropped and written off because they are associated with failure.

The confectionary industry is worth billions and the demand for chocolate products continues to increase. This industry too follows evolutionary principles. The competition between companies is intense, which drives new creations. The successful chocolate bars are retained because they sell, but they undergo subtle changes as the law demands. The wrapper is changed to meet a changing market-place. The sugar content is lowered to counter the problem of obesity. The bar itself is reduced to help keep the price stable due to higher labour costs or the higher price of coffee beans. The hunt for new products continues to keep ahead of the competition. From time to time, amid great fanfares and advertising, a new product is announced. Celebrities are paid to advertise the new chocolate bar and endorse it as the eighth wonder of the world. If it is successful, the new bar is added to the company's portfolio, which in turn increases the bottom line. If it is unsuccessful, the new bar is rendered extinct and not mentioned again. Nobody wants to be reminded of failure.

The company that desires to be successful and pass down that success to the next generation must imitate the laws of evolution by continuous innovation and experimentation. It must continue to try out new products, just as Mother Nature tries out new life forms. It cannot stand still but move forward, just as evolution never stands still. If it is to survive the cut-throat business environment, it cannot afford to remain unchanged or it will become obsolete. The market now plays the role of the predator. The company must continue to research and produce new products, or risk losing out to a more innovative rival. It can therefore be seen from these two examples that modern man imitates Mother Nature in the business field, although he is probably blissfully unaware of her subtle influence.

An analogy can be made here to a child growing up in a home with his mother. He leaves home in adulthood to live an independent life, believing that he has left his mother behind. He is unable to distinguish between his *physical* departure and his *psychological* departure. His body has moved away from home, but he carries his early life in his mind as he journeys into adulthood. Her influence continues to shape his life, although he is now an adult who can think for himself. His actions and thoughts are shaped by his formative years at home. His mother's subliminal influence remains with him until the day he dies.

Mother Nature is his original matriarch, the primeval giver of birth. She is mother to all life forms, of which man is but one of countless. He is not aware that she governs his life because he is removed from his roots in the natural world, and believes his destiny lies within his own grasp. He obeys the four immutable laws, without realising how he is profoundly influenced by them. He obeys the first law of nature by struggling for life. In sickness, he attends the doctor and fights against death with every possible means at his disposal.

He obeys her second law of establishing a territory by securing a place to live. His apartment or his house becomes that territory and he will defend it against attack, even from his own government. He is conditioned by the second law to stand his ground and protect his home. He obeys her third law when he marries and passes down his successful genes. If he has children, they will be different under the process of *neotento,* each child a new experiment in the laboratory of life. His sons shall be different to his each other, and his daughters shall be different to each other. He competes for work against his fellow men and women, and seeks to match their success in the promotion stakes. When he completes his life cycle and dies, his home, or territory, is inherited by his children. They have been given the best start in life by inheriting his property. His life has been just one of countless millions that have passed this way, in adherence to the four laws of Mother Nature.

IN OBEDIENCE TO NATURE'S LAWS

Mother Nature rules on Earth and all life forms obey her four laws which are not discretionary but compulsory, and are not open to change nor challenge. The deities that man invented also handed down laws, but they have not lasted because they were not eternal. Neither did the secular laws drafted by kings or governments stand the eternal test of time, and were either replaced or modified in line with the changes in society as mankind became more civilized. The laws that man ascribed to deities, and claimed to have received from gods, have too been superseded by the passage of time. These laws are now only used in theocracies that are doomed to failure because they contravene the natural laws. The stoning to death for adultery that is required under religious law has obviously no place in the modern era, and should be consigned to the dustbin of a more barbaric age. This religious law is now only retained by the most primitive and backward societies where human rights are absent and where religion holds sway over the minds of the people.

The secular laws of man have had to evolve to meet the changes in society, and continue to change as they must. Laws drawn up five centuries ago cannot possibly work in the modern era because they belong to another and darker age, and were framed to reflect the thinking of that time. In stark contrast, the laws of Mother Nature apply to all life forms on the planet, and do not change nor cannot change because they are endemic. These four laws are timeless and immortal; and have remained constant and true since life first appeared on the planet. Man might like to believe he is a creation apart, that he is above nature, but he too obeys her laws. It was not the laws that mankind ascribed to supernatural beings in the sky that aided his survival from a primitive past to a civilized present. Neither did the secular laws of kings ensure that survival and eventual mastery of the planet. The immutable and timeless laws of Mother Nature are responsible for his success. Man could not have survived without obeying the natural laws, and he could not have evolved to his present state of mental and physical development without them. The first law of survival ensured that his species did not die out, but lived and became dominant.

The second law of territory can be observed in the way the human race lives today. Nation states exist behind fixed territories that are clearly marked out with national flags and other signs of ownership that are universally understood and recognized. This law was not conceived in the mind of mankind, nor handed down from perceived deities in the sky. It is instead a fundamental law of nature that governs how territorial species must live. The third natural law can be seen in the escalation of the world population which currently stands at around seven billion people or more, and continues to rise.

The fourth law determines how the world lives in the third millennium CE. There is competition between individuals, between peoples, and between nations. That competition underwrites trade as companies compete with each other. It governs how individuals compete with each other in the marketplace for jobs. It decides how a nation reacts to a threat from a rival. Any advantage gained by that rival is countered by a similar or superior reaction from the nation under threat, as demanded by the fourth law.

Man has reached this stage of his evolution in obedience to the four natural laws and not to any others. To illustrate the imperative of natural laws, and also to examine what happens when man interferes with the work of Mother Nature, it shall be necessary to look at invasive species. Where species evolve in the same environment, predator and prey maintain the essential balance that underpins the natural world. But where man interferes in the natural order, chaos is the usual result. Japanese knotweed was introduced from its homeland to Britain about two centuries ago. It had natural predators in Japan that kept it in check, but not in its new home where it found the conditions ideal and spread like wildfire. The plant uses its root system to take over territory, and they are so pervasive that they can undermine walls and bring down houses. Property prices drop when this invader is found in the garden, and it takes trained specialists and constant work to remove and destroy it.

Something similar happened when rhododendron was also introduced. It too began to eliminate the native competition and needed man to put right the damage he had caused in the first instance. The plant is now being cleared from nature reserves and parklands where it was out of control and threatening native species. Both rhododendron and knotweed are invasive species that obey the natural laws, and prove that they govern both fauna and flora alike.

The grey squirrel was introduced as a pet, but escaped from captivity into the wild. It outcompeted the native red squirrel and wiped it out in many forests. The grey also carries a pox that the red has no immunity against. Once again, man was required to put right his error by establishing safe havens for the native species to ensure their survival. The invasive species, the rhododendron and the grey squirrel, obeyed the four natural laws that Mother Nature put in place. They survived, they established territory, they passed on their successful genes, and they outcompeted the competition. By adapting to their new territories, they improved their genes and passed them down to the next generation.

The inevitable result of man's interference in the laws of nature requires him to rectify his errors, and repair the damage caused by his basic lack of understanding for the natural world. The example of the grey squirrel outcompeting the red has echoes in the European invasions of the American continent. The natives had no natural defences against the diseases that the invaders introduced, and in some instances ninety percent of the indigenous population was wiped out where they came into contact with the European conquistadors. The role of disease shall now be examined as part of nature's great plan to improve the species.

Disease plays an important role in the health of a species, and that role applies to flora and fauna alike since both are governed by the imperative to improve. No disease evolves to kill one hundred per cent of a species or population, but to strengthen its core by testing it in an attack. The individuals who survive are made stronger and pass on their successful genes thereby rendering the next generation more robust, but the weakest are killed off in the attack and their genes eliminated. Animals and plants undergo similar trials designed by Mother Nature to weed out the sick and improve the strong, and help the next generation by granting it immunity.

Disease should not be seen as a random event, but part of the natural order to ensure that only the best genes are passed down by eliminating the weak genes. It should be regarded in a similar manner to the periodic droughts on African savannahs that test flora and fauna for fitness. The species that survive the droughts pass on their successful genes, and the unsuccessful species that do not survive the drought gave their genes extinguished. Disease plays an essential role in the health of a species, just as a predator plays a similar role. As a result, diseases can never be eradicated because Mother Nature designed them to improve life.

Probably the greatest pandemic in world history was the Black Death that occurred in the fourteenth century, and the fear it generated still sends chills down the spine to the present day. Estimates vary on the numbers of people it killed, ranging from a third of the European population to fifty percent. The survivors were made stronger by the disease, and passed down immunity in their successful genes. Epidemics and disease should therefore be regarded in the same manner as natural predators, constant companions of mankind to improve the species. If diseases were not essential tools in improving the species, Mother Nature would not have evolved them.

In 1918, a flu pandemic swept around the world and killed approximately five per cent of its population. It was called the Spanish flu although it did not originate in that country. Historians now believe that it arrived in the baggage train of American soldiers who travelled to fight in World War One. It was first reported by the Spanish press, hence the name, but no disease has a nationality. The Spanish flu targeted the young, the old, and the sick. These are exactly the same segments of populations that predators target, those with the weakest genes. The timing of the pandemic is no coincidence either, coming right at the end of World War One, the most devastating war in history up to that time. A pandemic, just like the predator, is an opportunist and always strikes when the victim is at its weakest. The food shortages and lack of medicines caused by the war weakened populations and left them more vulnerable to attack. The purpose of a pandemic is not to wipe out populations as a whole, and this is the reason why there are always survivors, but to test it for weakness. The weakest are eliminated, but the survivors gain a level of immunity which is then passed down to their offspring.

During the middle of the twentieth century, rabbit populations exploded causing widespread damage to crops, followed by demands from farmers to control their escalating numbers. Man came up with a virus called myxomatosis to wipe them out and solve the problem. When it comes to wiping out species, no animal can compete with man. The virus was particularly virulent and destructive, but it did not wipe out the whole population. No virus has ever completed that task because the laws of nature do not permit it. Some rabbits survived to breed as only rabbits can, and the next generation gained an immunity from the virus.

The virus did not get rid of the rabbit problem, but instead improved it by making the next generation resistant. That is essentially the role a virus or disease performs. The laws of nature had foiled the best efforts of man do deal with a problem he himself had caused by growing monoculture crops that allowed rabbits to increase their numbers. Man had also helped the rabbit population explode by using pesticides that entered the food chain, and reduced the ability of predators like falcons and hawks to successfully rear their young. When the balance of nature is upset, the result is chaos.

As man continues to battle epidemics and disease, they will adapt and change to meet the new challenge under the fourth law. Stronger medicines will be met with more resistance from the disease as its immunity to them is built up in each generation. The arms race between man and disease will continue as he vainly seeks to gain an advantage over the disease, and then the disease will counter by matching that advantage. This is precisely how the fourth law of nature works. An advantage by one side is always countered by the other side.

Man has been battling epidemics and disease for millennia, and has an abysmal record. Despite the vast sums of money and innumerable manhours he has expended in the war, he has managed to eradicate only one disease and that is smallpox. He does not seem to realise he's engaged in a war he cannot win. He might win an occasional battle, as with smallpox, but the war is unwinnable. He is engaged against the immutable laws of Mother Nature and fighting a lost cause. He can spend billions of dollars in an attempt to cure the common cold, but he cannot defeat it no matter how much he spends. The common cold operates under the fourth law, and keeps adapting to counter each new medicine that comes along. Instead of examining disease through the lens of a microscope, doctors and scientists should examine it through the clearer prism of nature. As already pointed out, the proper study of mankind can be found in the natural world.

The disease survives by fighting for its life, as required under the first law of nature. It establishes territory in its host by obeying the second natural law. It passes down its genes by multiplying within its host, and acquires new territory by infecting more hosts. When a medicine is used to combat the disease, it competes against the drug as required under the fourth law. One example of many can be found in the current war against cancer. It continues to defeat the best efforts of the medical profession to eliminate it, which proves how long and difficult the war has been. Cancer is just one arrow in the quiver of diseases introduced by Mother Nature in her process of eugenics to improve a species. By divorcing himself from Mother Nature, man has lost contact with her laws and no longer understands how they work. If a cure for cancer is found, then another and equally terrifying disease will evolve to take its place. Nature abhors a vacuum, and always fills it.

The latest pandemic to sweep the world is Covid-19, and it has almost brought the planet to a stuttering halt. The helplessness of man when confronted with nature has been amply demonstrated by the latest pandemic. This is a novel virus, meaning it is new and therefore unknown and with no previous record. It would appear that nature is one step ahead of man yet again in the continuing war without end, in which there can be but one winner. All diseases and pandemics follow the natural laws because they too are part of the natural world. Under the third law, successful species are required to pass down their genes, and they employ different strategies in this effort.

Cedar and ash trees, for example, use the wind to carry their seeds and pollinate the next generation. Fruit trees such as apple and pear use go-betweens like bees and other insects to carry their seeds and ensure the continuation of their genes. This is the same method that the Black Death used to spread its genes, by using a flea that jumped from rats and infected the human host. Mammals must have tactile interaction to pass down their genes during the act of copulation. This was how the HIV virus spread with sexual activity between humans. It infected the homosexual community first before jumping to the heterosexual community. This disease is still out there and killing countless numbers of people across the world.

Mankind has never faced an enemy quite like disease in its long and bloody history of warfare. Some diseases are protean, and can change and morph into something new and unforeseen. Others can act as Fifth Columnists and hide within the host, unknown until the time to strike arrives and then it shows itself. Some can secretly infect the host, who shows no symptoms and is unaware of the infection. That host then infects others, almost acting as a recruiting agent to spread it like wildfire. A disease can also jump species. Swine flu and chicken flu can infect humans, and Covid-19 is believed to have originated in bats or pangolins.

This latest pandemic seems to be an evolutionary leap forward by using multiple methods to find a host. It can use the air to find a host, just as the ash and cedar tree. Droplets from an infected person are carried on the air in a sneeze or a cough and can find many hosts who inhale them whilst breathing. It can distribute its genes by touch, and human beings are a gregarious species that like to live in close proximity. It can infect surfaces and worktops where it remains unseen until finding a host who picks it up, without realising the danger. This is the enemy that mankind is fighting against, an implacable agent of Mother Nature that has been a constant companion of man since he first evolved.

The war shall be long, but no victory is possible.

THE RELEGATION OF SMELL

Mankind is classified by science as part of the primate family which includes monkeys, gorillas, baboons, and chimpanzees. Each group within the extended family shares a common law that places a dominant male as the leader, and he sires most of the offspring. Although some males use cunning to breed with females, they risk a beating if caught by the dominant male. Studies have shown that the alpha male still sires more than fifty percent of the offspring, as the law of nature demands to ensure the best genes are passed down.

Chinese emperors and Ottoman sultans had harems, a legacy of their primate origins that required the alpha male to sire the next generation. It was essential to guard their harems to prevent lower-ranking males from having sex with the ladies, and thereby perpetuating weaker genes which is contrary to the laws of nature. The leaders solved this tricky problem by removing the testicles of the guards and creating eunuchs, and in the process ensuring that only the strongest genes were passed down. Emperors and sultans alike were obeying the third law of nature by guaranteeing only the best genes were passed down by employing guards who could not have sexual relations with the harem, for obvious reasons that require no further elaboration. The message that only the leader must have access to the harem was a subliminal reminder of that past age when the dominant alpha male sired the next generation, and weaker males were excluded from the breeding cycle.

Primates are territorial animals and congregate together in troops, living in trees or on the ground. Each troop marks out its respective territory in a different way. Not all sub-species on the family tree evolved at the same rate, and consequently some are more advanced than others. The lower orders, such as tamarind monkeys and lemurs, mark out their territories by scent, which they use on tree branches to advertise that their patch of the forest is already occupied. These are olfactory warnings to rivals that the territory has a resident troop, and they act as a deterrent to warn off trespassers. The scent-marked territory is an olfactory warning that it shall be stoutly defended against intrusion by its present owners who have claimed it as their own property.

Chimpanzees are a higher order of primates, and they abandoned scent-marking in favour of visual signals. This profound change probably occurred when the more advanced apes abandoned the trees in favour of living on the ground. Tails were also abandoned under the evolutionary law of use it or lose it because they were no longer needed on the ground. Tails were useful in the trees for balance, and they could also be used to grasp branches like another hand. They were superfluous to ground living and as a result became redundant. Just as the muscle that is not used atrophies and eventually dies, likewise the tail that was not used eventually atrophied and died.

Chimpanzees do not scent-mark and they are the closest relatives to humans. Their territory is defined by displays of strength at its borders, accompanied by tree-stomping. In this manner, visual signals of ownership are accompanied by auditory signals, sending out a double message of ownership. This was a profound change in territorial behaviour by the higher primates, and inherited by the branch that that evolved into humans that is now known as Homo Sapiens. Man followed in the footsteps of his chimpanzee cousins to map out and establish his territory, using visual signals to declare ownership. Scent was relegated and abandoned, to be replaced by sight as the most important sense. This move would further alienate man from nature, and set humans on course to dominate the world.

The resultant loss in importance of the olfactory sense led to the relegation of smell under the law of use it or lose it. This sense would further decline in importance as man continued to climb the evolutionary ladder. Animals that still use scent to mark out territory can identify rival members by scent alone, as any owner of a domestic dog can attest when he takes it for a walk. The dog sniffs lamp posts and car wheels to discover who has been in its territory, and can probably identify the culprit by his scent. Man has lost the ability to determine who has been in his territory because evolution relegated the sense of smell in his ancient ancestors, and promoted the sense of sight to the foremost position.

Eyes replaced the nose as the primary sense to identify if a territory was occupied by a rival tribe. Now sight became by far the most important sense, and modern humans too recognize that it is the foremost sense. The loss of any other sense is not as debilitating or has a more lasting impact than the loss of sight, a measure of its vital importance. Humans take it for granted until sight is impaired or lost, and only then realise its vital importance. Living without sight is to dwell in a place of permanent darkness where light cannot penetrate, and where territorial markings are not seen.

Modern man has inherited the system of clearly marking out territory by visual displays, and it continues down to the present era. Dykes, walls, and earthworks were erected and used as visual warnings to signal ownership of a territory. If the signs were ignored and a successful invasion mounted, the victor would display his own visual signs. Invasions of a rival's territory were always accompanied by visual signals of new ownership, and still are. Where Norman conquerors invaded and won new territories, their declaration of ownership was the stone castle in a prominent place. These were visual signs that the conquered lands had new and powerful owners. The castles were built on high positions, not only for easier defence but also that they could be seen for miles. Nobody with a pair of eyes was left in any doubt about who owned the territory.

Now, national flags at borders send out a visual signal that the territory is occupied. Would-be aggressors are left in little doubt that the territory is occupied and will be defended. National anthems reinforce the same message by sending out auditory warnings. The policy of using both visual and auditory signals to spread the message that the territory is already occupied was inherited from the chimpanzee branch of the family tree. These twin messages are now universal and apply to every country in the world. The visual warning of a national flag that modern man uses to declare ownership is backed by an auditory warning of a national anthem. The territorial warnings have survived down to the present era from a primitive age, proving so effective as a deterrent and a declaration of ownership that they have been retained by modern man. The twin warning threats are clear and universally understood, and leave potential aggressors in no doubt about a response. They signal that the territory is already occupied, and shall be stoutly defended against any attempt at a takeover. Both sides know the rules because they both belong to the same species.

Visual signals are so powerful that they now dominate the world of commerce, and a logo can inform a potential client about everything he or she wishes to know about a certain product. As the ancient Chinese proverb puts it, a picture is worth a thousand words. A logo can inform the prospective customer where the product was made, whether or not it has a good name, and if it is reliable. Global brands do not need interpreters to sell in foreign markets with a different tongue, the logo does it for them with no language barrier. The eyes of the customers act as the interpreter, just as they identify the flag of a powerful country.

The profound change from the outdated sense of smell to the more efficient sense of sight undoubtedly spurred the further evolution of the brain in mankind. Sight encompasses a far greater range of sensations than smell, and one can be mentioned to prove this point. Smell cannot distinguish colour, whereas sight can not only identify a wide spectrum but can also isolate and store each one independently in the brain. Each new visual sensation created another neuron connection, which advanced the brain's capacity to acquire more knowledge. And the brain that receives more information becomes more intelligent. That increased usage and storage of new information stimulated its evolution.

Sight also provided a better defence mechanism for early man than a sense of smell. A predator, by staying downwind, could creep up on a victim almost unobserved. Eyesight is not affected by wind direction, and standing on two legs gave man a clear field of vision on the open plain, allowing him to take cover or defend himself. He could identify a threat long before it became a danger to his survival too. Furthermore, the dangerous animals were identified and stored in his brain, allowing him to avoid them in future.

Another change in the behaviour of chimpanzees that resulted in the change from smell to sight would also affect human sexuality. Scent plays an essential role in mammals to sire the next generation that has remained unchanged long before man emerged on the planet. When the female in many species come into oestrus, hormonal changes in her body inform males that she is ready to mate. These changes can be detected by the male who sniffs or tastes her vulva or urine to determine if she is in season. The male lion has a scent gland in the roof of his mouth called the Jacobsen's organ which enables the animal to determine if the female is in oestrus, and therefore ready to mate. A female leopard advertises her availability by spraying her scent on a tree, which can be detected by a nomadic male.

The higher primates, by abandoning the sense of smell in favour of the sense of sight, had to solve the problem of knowing when females came into oestrus, since their noses could not tell them when the females were in heat. Now the females had to visually display when they were available because eyes had replaced the nose to determine when the female was ready to mate. These visual signals can still be observed in primates today. Female gelada baboons that live in the Ethiopian mountains highlight their availability by red swellings on their chest and genitalia. Likewise, the genitalia of female chimpanzees turn reddish pink and swell when they are in season and ready to mate. The period of fertility lasts about thirty-six days, and they only conceive during this very brief window of opportunity.

The profound change from the olfactory to the visual sense required the females to evolve to meet the challenge. Since the male could no longer identify the female in oestrus by scent, he was now required to identify her by sight. His eyes now had to perform the task that his ancestors had done with their nose to perpetuate the species. Proofs of this change can be found in modern sexuality between consenting men and women.

Oral sex between modern couples is practised by both sexes and not regarded as abnormal or out of the ordinary. It plays no part in actual procreation, but rather conducted because it adds pleasure to the love-making. There is an explanation for every facet of human behaviour to be found in nature, and the reason for oral sex is a legacy from the past. Before the higher primates evolved from scent to sight, they had to discover if the female was ready to mate by scenting her urine or tasting her vulva. Just as bullying in the schoolyard is a throwback to an era when the strongest ruled, oral sex is a legacy of an era when scent and taste were essential to discover if the female was ready to mate. Just as the coccyx bone is a reminder of a lost tail, oral sex is a past reminder of the loss of scent to identify oestrus.

The sexual evolution would continue as early man further distanced himself from nature. His ancestors had lived in a troop dominated by a strong alpha male who sired the next offspring by mating with all the females. The natural imperative to mate was so powerful that another male risked a beating from the alpha male by mating with a member of the harem. This would have led to instability within the troop and threatened its vital cohesion. A divided troop is more vulnerable from attack than a united troop. Another system was needed to allow males to mate without fracturing the vital cohesion of the troop, and jeopardise its very survival. The problem of avoiding strife over access to females was solved by pair-formation, and again females adapted to meet this behavioural change.

Chimpanzees and gorillas still live in the wild in troops dominated by an alpha male who sires most of the offspring, but modern man lives in a monogamous relationship. Man is the only higher primate who has abandoned the harem system in favour of pair-formation. Gorillas and chimpanzees live as they always have and have not changed over hundreds of thousands of years, but man has evolved to form a bond with a single member of the opposite sex. That evolutionary advance was a profound change because it allowed for the development of human civilization, and marked a further departure from nature. By abandoning the tribe to form a personal relationship with a member of the opposite sex, this most important step gave birth to individual thought, which would eventually lead to the discovery of philosophy and other disciplines of the mind. Man learned to think for himself outside the influence of the tribe because he had to fend for himself and care for his family.

Although pair-formation is how the vast majority of people across the world choose to live, mankind is not naturally monogamous. The evidence can be seen in the extra-marital affairs conducted by both sexes. They can promise on their wedding day to be faithful to each other, but some are hearing a siren voice from another era. The past often intrudes on the present, and these affairs are a throwback to the distant past when humans lived in a troop dominated by the alpha male who had a harem. Then lower ranked males secretly mated with the alpha male's harem when they got the chance. Females also benefitted from these covert liaisons because it increased the chances of conception, and mixed the gene pool.

The driving force of sex is the perpetuation of genes, and the female who could increase her chances of conception by mating outside the harem took the opportunity. Her role was to produce the next generation, and to increase her chances of pregnancy. The modern married man or married woman who has an affair outside marriage risk losing home and family, but that does not stop them from having affairs. The siren call of the past echoes down to the present, and this legacy of evolution cannot be easily abandoned. The married man who has an illicit affair has his counterpart in the troop, who secretly mated with a member of the harem when the alpha male's back was turned.

The married woman who has an affair has her counterpart in the harem, who covertly mated outside the harem. How pair-formation came about is unknown of course, but it probably occurred when man gave up his nomadic lifestyle and settled down to farm. Living in a settled community with a ready food source allowed a couple to store supplies during the winter months and to tide them over droughts and bad weather conditions. Farming and herding their own food supply gave their offspring a better chance of survival. They had a ready food source on their doorstep that ensured the children would not grow hungry, thereby increasing their chances of survival.

The discovery of farming was another step on man's way to the top of the evolutionary ladder and would also mark a further departure from his simian roots. He had not only split with nature, but also from the primate lifestyle and method of procreation. The change from living in a troop to a monogamous relationship required the woman to adapt her sexuality. She changed from the oestrus cycle to the menstrual cycle, which helped bond one man to one woman in a monogamous pairing. The troop was left behind in the evolutionary race and abandoned in favour of pair-formation. Each step up the evolutionary ladder further distanced man from his original roots.

The menstrual cycle marked a profound change in the evolution of humans that marked a fundamental change from nature. This arrangement can be regarded as the birth of civilization. It could not have occurred under the previous system when they lived in a troop that acted and thought collectively. Pair-formation between one man and one woman was the foundation stone on which the monuments of civilization were first built, and paved the road ahead for humanity.

Now instead of coming into season occasionally, the woman was always available and could conceive all year round. She was not confined to her season, and she was not restricted to a brief window of opportunity when she could conceive. Each month an egg was released in the lunar cycle that could be fertilized by her partner. As a result, she had no need to advertise that she was ready to mate, unlike her primate cousins. Furthermore, she was paired off with her own male who had a personal stake in her offspring because he had fathered them. He had a personal stake in his offspring because they were bearing his genes. He had an added incentive to protect them and rear them into adulthood because he was protecting his investment in the future.

This was a far better arrangement for both men and women than living in a harem with an alpha male who could be deposed at any time. Teaming up offered better protection for the mother because her partner had a vested interest in her children since he had fathered them, and was therefore incentivised to stick by her side until they were reared to maturity. It was a better arrangement for the father too since he knew that the children were the offspring of his loins, unlike the harem system when nobody knew who had fathered the children. Pair-formation solved that confusion and gave each couple a shared interest in raising the children. The cohesion of the troop or the tribe was better guaranteed under this evolution too because there was now no need for covert sex, and the mayhem that ensued if the perpetrator was uncovered by the alpha male.

It can reasonably be assumed that that evolution produced a special hormone or chemical to bind one man to one woman, and vice versa. That hormone, chemical, or bonding agent must remain unknown until future science can identify it, but it does exist. Love is a very potent emotion and can be stronger than even death. Men and women have given up their lives so that their loved ones could live. Self-preservation is the first law of nature, and to abandon it for the object of one's love is surely the supreme sacrifice anyone can make.

Vestiges of a distant era where the alpha male had a harem still exist in some societies especially those with long traditions of the practice, and of course religions that are not susceptible to change nor moral censure. But the overwhelming majority of people across the world live in marriages based on one man and one woman. The relegation of smell in man's earliest ancestors and the promotion of sight was the initial catalyst for this change to come about. The love between one man and one woman would lead eventually to great works of art and literature. The pursuit of art and literature in their various forms would further stimulate the evolution of the brain. Art and literature are conceptual pursuits, and artists and writers form a mental picture of their work before taking up a brush or a pen. Humans are the only creatures that create art for pleasure, and they have the most advanced brains on the planet.

Science is now coming up with proofs that some men can detect when a woman is most fertile during her menstrual cycle. Since women do not exhibit outwardly visible signs of their optimal time to become pregnant, the conclusion must be that these men can pick up the hormonal changes in her body by smell. If evolution is considered, this recent discovery makes perfect sense. Apparently, some men retain a latent sense of smell that can identify the hormones that signal the woman is in her most fertile period. That was how their primate ancestors discovered whether or not a female was in oestrus after all, by detecting when she was ready to mate by smelling hormonal changes in her body. Although relegated to the point of irrelevance, it would appear that a few individuals still retain the ability to detect hormones by their olfactory sense.

The past never really goes away.

THE DICTATES OF LIFE

If it is true that we live in a man's world, how did this present situation come about? Why has war been a male preserve since man's ancestors first took the branch of a tree and used it as a club to defend himself or attack his neighbour? There is an explanation for all facets of human nature, and both these questions have origins in the primate system of rule that governs mankind down to the present day. That society is founded on physical strength, and since males are physically stronger than females they are the dominant sex. This is also the reason why the leader of a tribe or troop must be an alpha male, not an alpha female. The system as prescribed and laid down by Mother Nature precludes females from the top job for this reason.

Mankind is descended from a simian society dominated by an alpha male who relied on strength to maintain the cohesion of the troop. He was the best fighter who led the troop in wars of defence and in wars of territorial expansion, and they rallied around him. The male was better equipped for fighting than the female because of his superior strength and ability, and the role that nature had decreed. He had to fight to retain his position as the alpha male, and to protect his troop from rival troops. The human male is physically stronger than the female, and that position did not come about by chance but by evolution. Men go to war because they are better adapted for fighting, just as the males in the troop went to war in defence of territory or its expansion. The roles of the sexes were clearly marked out and defined, with males as fighters and females as mothers and carers.

Here is the reason why men have traditionally ruled and not women. Mother Nature had set down the rules for each sex to follow and these rules were infrangible. Primate society forbade females from the top job because they could not fight and therefore could not protect the troop, the most important requirement in a leader. Protection of the troop was the essential and by far the most important qualification for the role of alpha male, or troop leader. That role is still the most important in the modern world.

Mother Nature had not equipped females for that essential role the leader must perform, to protect the troop or the tribe. The reason why it has taken women so long to become presidents or leaders of a country has nothing to do with perceived male bias nor discrimination against women, but with this primate legacy. The primate system of rule demands a strong alpha male, and by definition a female is excluded from that position. The subliminal message from the past that a male must rule is part echoed down to the present era, hence the current leadership roles in the modern era, which are predominantly held by men.

Human politics did not come into existence by the throw of a dice or serendipity, nor were they handed down from any higher being who lives in the sky. Modern systems of rule are solely derived from primate ancestry, and particularly from chimpanzee society. Man inherited the system from this branch of the primate family and has modified it to suit his own lifestyle, but still basically it remains true to the prescribed model of rule. The alpha leader who headed the earliest troop did not reach the top by brute strength alone, but was aided by that most important organ between his ears. He used his brain to cultivate allies and form a gang of enablers. The modern candidate who now wants the alpha male job does not have to be the strongest, but he does need to be the most cunning and the best at cultivating allies.

The basic model of primate rule is an alpha male leader who is assisted by enablers or allies who help him achieve power. In return, he rewards them with perks and protection. Every system of rule on the planet is governed by this model, and it cannot change because that is how the species evolved. The world too is governed by the fundamental primate model and has always been since the dawn of human time.

In the twentieth century, a Cold War rivalry between the USA and the Soviet Union threatened to plunge the world into a nuclear war where there could be no winners, and the world held its breath as both sides competed for the alpha male role. Man is seldom rational when it boils down to war, and for a time humanity stood on the brink of destruction. The Cold War almost turned hot when the Soviet Union placed nuclear warheads on the island of Cuba and the USA threatened to retaliate with its own nuclear arsenal directly against its foe. The primate system was at play, except few if any understood what was happening because modern man had lost his way in the split from Mother Nature.

The USA was the alpha male in the West, and the Soviet Union was the alpha male in the East. They can be compared to two rival troops sharing the same forest but not the same territory, each having its own boundaries with the demarcation lines clearly marked out. The Western allies of the USA were the gang who supported the country in the struggle against the rival. In return for this help, the USA played the part of the alpha male and protector of its allies. If one of the member states faced an attack by the Soviet Union, the USA would come to the aid and defend it from the enemy. That was its function under the role of the alpha male in primate society.

The Soviet Union played the same role in the East. That country also had a gang to support it in the struggle against the foe. If any of its allies were attacked by the USA, the Soviets would come to their aid. It was not common sense that prevailed in the end to restore sanity, but the weakness of one alpha male in the East that could not afford to match the financial clout of the West. With the collapse of the Soviet Union, members of the former gang sought a stronger alpha male to protect them. Poland was once allied with the former Soviet Union, but after the collapse of that union the country has become part of NATO. Now that organization is the strong alpha male that gives Poland the vital reassurance of protection. This is the reality of competition between the same species as ordained by Mother Nature: they who are unable to match their opponent, or allow that rival to gain an advantage, go under.

Trust is vital in the relationship between the alpha male and the gang or allies, and it cuts both ways. Each party must be certain that the other will come to its assistance in times of crisis and offer any help it can. If that solemn pledge is broken once, it can seldom if ever be re-established because an ally that lets you down once cannot be trusted when the next crisis looms. There is a greater obligation on the alpha male than on any member of the gang, because one member of the alliance can renege on his promise of aid in times of crisis; but the leader must fulfil his vital role or risk losing the whole gang. Worse than losing face, he can never persuade another gang or allies to accept his leadership and follow him again because he is not to be trusted.

A promise of aid must be kept or he will stand alone in future, and the world is a dangerous place where allies are essential for a sense of security. And his lack of action shows him up as weak both at home and abroad, and emboldens his enemies to act. Weakness in a leader is never forgiven, and abandoning an ally is an unforgivable stain on a nation and on an alliance. That is a stain no amount of white-washing can erase.

The civil war in Syria follows the same brutality that only this type of civil conflict produces and always has in the past. Brother against brother and son against father is invariably more barbarous and crueller than any conflict between strangers, because both sides in the conflict are fighting for the same territory that they rightfully consider their own. This civil war is more brutal than many because chemical weapons against civilians are being used to clear areas of rebels and retake territory.

War is never about people, but always about territory. The images of children suffocating and unable to breathe after chemical attacks stirred outrage around the world. The lack of medicines to cure the sick and injured made these haunting images almost unbearable to watch. Sadly, these images, in the usual ways of the world, have been transitory and are now largely forgotten. These were merely a sideshow to the main event which was the battle for territory. That is more important than human lives and always has been, even if the world refuses to acknowledge the fact or recognize the truth. The world has always preferred platitude to truth, and always shall because the latter is too uncomfortable to consider.

Children unable to breathe after a poison gas attack is old news now and is now not mentioned. Distraught mothers pulling their dead children from rubble no longer merit a line of sympathy from the outside world. These images have faded from the collective memory, and have been forgotten. The world has moved on as it always does, and these dreadful snippets of civil war belong to the past. People do not want to remember atrocities, and so they have forgotten the terrible images of civil war. What will not be forgotten but remembered is the role of Russia in the conflict. That country performed the role of the alpha male by coming to the aid of his ally, the president of Syria. That salient fact, and that fact alone, will be long remembered when every other brutal image has vanished from the collective memory of the world.

Where the alpha male partner does not fulfil his role as protector, that vacant situation is invariably filled with his rival who does fulfil the role, and carries out his responsibilities. The new alpha male partner then attracts others into his orbit who also seek his protection, because he has demonstrated his reliability by coming to the aid of an ally in distress. They see how he has reacted to a threat to one of his gang, and are suitably impressed. He has demonstrated that he shall come to their aid too in times of peril and fight for them as promised. That basic and primary need for protection is part of primate origins, and has been since members lived in a troop protected by a strong leader. That need is still hardwired into the human psyche and cannot be erased because it concerns survival itself. The alpha male leader that serves this need will have many allies to support him. Conversely, the protector who fails to come to the aid of a member of the gang is soon short of allies.

As tribes became larger and territories were expanded to encompass conquered peoples, the alpha leader needed a new system to build cohesion and to secure the future of his offspring. Kings evolved to fill the vacuum, thereby creating a ruling dynasty with hereditary powers. Now there would be no need for the selection of a new alpha male on the death of the incumbent since he had a son ready to wear the crown. That crown and that royal title commanded the support of the people, and their loyalty kept new king in charge at the top.

As territories grew larger still and disparate regions and tribes taken over, the title of king evolved into the title of emperor. An emperor is more important than a king because the title itself conveys the message that he controls more territory. That is the imperative of the second natural law, the acquisition of more territory and the reason why the empire evolved. Empires are compared by the amount of territory they conquer, and seldom whether they were good nor bad or how they treated the peoples under their rule. The vital importance of territory to humans overrides most other considerations, and always has.

The evolution of the alpha male in the past has resulted in the politics of the present. Kings do still exist, but mainly in ceremonial roles with no real powers, at least in Western societies. The emperors that arose and fell have entered the pages of history and their empires with them. The present alpha male is now the man with his hand on power, and that role belongs to the man with red blood in his veins and not blue blood. He could be an army officer who seizes power, a dictator who murders his way to the top, or a democratically elected leader. Whoever he is, he needs to have a gang at his side to back him up. That is the basic system of primate rule and it does not nor cannot change.

Army officers and dictators who seize power are always male; but a woman can gain the alpha male position in a democracy through the ballot box. However, she too must fulfil the role that's required of her because the obligations and duties of the job have not really changed in three hundred and fifty thousand years. The dictates of primate life do not change simply because humankind has left the past behind and become more sophisticated and advanced since its humble beginning. The female leader must perform the designated role of the alpha male by keeping the people safe. That is her primary function, and leading a democracy does not alter that vital and overriding obligation.

THE TANDEM EVOLUTION

The use of comparisons to the animal kingdom in this work is repetitive and quite deliberate. They are a reminder that man too is part of that kingdom, and that once he lived in harmony with the natural world. Although modern human beings still live by the four natural laws, they do not question how they were formed, where they came from, nor who laid them down. People believe that they were created by an omnipotent being in the sky called God, or they believe that they evolved by accident. The objective of this work is not to question the origins of mankind, but to positively and clearly demonstrate its close connections to the natural world which many people consistently refuse to concede. Modern society has become so sophisticated that it no longer regards Mother Nature as the original mother who gave birth to mankind, and fail to acknowledge that she continues to provide people with a home to rear their families, and materials to build their homes. They take the abundance with no thought of repayment.

Earth supplies materials for people to build cities and precious metals to make them rich. They never even consider giving thanks for the bounty that the planet provides, as ancient peoples did. They now believe that only primitive people far removed from civilization should contribute to such foolishness. Modern man has now become so distanced from his roots that he believes his split is complete, only to be reminded that his journey had a beginning in nature. He has reached the present stage of his evolution only by journeying from the past. He is on a continuous journey in evolution, and has now reached a stage in that unknown and unforeseeable future. Man could not have reached the present stage of that journey nor the present level of sophistication without passing through that early stage. He can be likened to a baby who must pass through the early stages of its development before reaching middle age, unsure if it shall ever reach the final stage.

The journey of man from a primitive past to a sophisticated present can be compared to the child who received a college education that his mother lacked. He does not bring his friends home because he's ashamed of his mother's ignorance. The educated young man has forgotten that his mother had worked to put him through college. That is how Mother Nature should be regarded, as the birth goddess and facilitator of mankind to reach its present level of evolutionary progress. Modern man, in his current state of sophistication, now believes that he is master of his own destiny, unaware that she pulls his daily strings as the puppet-master controls his creations dangling on strings.

Mother Nature is in control, not man.

Evolution is widely discussed in the physical sense alone, or in changes to the body as life forms adapted to meet the challenges of survival. What has not been examined fully are the changes in the brain that were also required, and they occurred in tandem. The brain had also to adapt to meet new challenges that evolution threw up. A change in habitat from dwelling in the trees to dwelling on the ground would have required a similar and equal change in the brain. The different types of environment would have required specialized skills and adaptations to meet the new challenges. The gradual loss of the tail by the ancestors of the chimpanzee required the brain to also adapt to the new environment. Living on the ground posed different challenges to living in the trees, and they had to be solved. The body could not have evolved without the tandem change in the brain, nor independently. Both body and brain had to advance in mutual tandem, working as a team to ensure survival.

An analogy can be made between two species that drove the evolution of each other, the cheetah and the Thomson's gazelle. The cheetah has faster speed but the gazelle has much better endurance, and can manoeuvre quicker. Each advantage gained by one had to be matched equally by the other. The prey and the predator evolved side by side, and thereby spurred the evolution of each other. The changes in the bodies of both predator and prey had to be matched by corresponding changes in the brain. Neither body could have evolved unless it was matched by a corresponding change in the brain. Each metre of extra speed required the brain to make an adjustment because a split-second delay might make the difference between life and death.

A chase by a cheetah can reach sixty miles per hour, and that speed demands an instant reaction to every move by the prey. Decisions are not made in the body, but in the brain. That organ calculates distance and relevant speed during the chase. It reacts to each turn and twist in the chase, and then makes the necessary adjustments. The relationship between predator and prey not only improves their bodies but also their brains. Predator and prey must react to any advantage gained by the other, both in body and in brain. The physical animal and the mental animal must advance together in tandem because survival depends on them working as a team.

The tandem evolutions that occur between predator and prey also occur in the individual; and in his journey from the past to the present man also went through the same phase. Another analogy can be used to describe how this happened. The young boy who receives his first bicycle must adjust from walking on two legs to cycling on two wheels. His body has to adapt a new posture, but his brain also has to cope with the new conditions and make the necessary calculations. The boy is almost certain to fall off until his body works in tandem with his brain, and they both work in harmony to ensure he can ride the bicycle in safety. As he grows into young adulthood, he acquires a motorbike and that presents more challenges in body and in brain. He must now adapt his body posture for the new challenge, and his brain must deal with dangers he has not met before and react if the speed of the machine is too high. His first car presents more challenges; but by his brain and body working in tandem he can meet them and progress into adulthood.

Within each species there is also tandem evolution between male and female. Their bodies are adapted to fulfil their roles in each particular society, and likewise their brains. An example can be seen in lion prides where the males are larger and stronger than the females. Their bodies are adapted for the roles they have to fulfil, such as fighting for the alpha male position, securing their territory, protecting the pride from external enemies, and siring the next generation. They have male bodies for these roles, but crucially they also have male brains. Mother Nature has equipped them with the body for their role in the pride, and with the brain to carry out their role.

Lionesses are smaller and lack the strength of males because it is not necessary for their assigned role in the pride. Their bodies are designed for bearing the next generation, not for fighting off rival prides that desire to take over their territory. They are the providers and the mothers of the next generation, and as such do the majority of the hunting and also suckle the young. They have female brains to help them fulfil their role in the pride. Male and female bodies are different because have been designed for their respective roles within the pride, and their brains are different for the same purpose.

Chimpanzee troops operate similarly as lion prides, with the males stronger than the females. They are better equipped for fighting because that is what their role demands, and their bodies have been designed by nature to perform that function in the troop. They have to fight for the top job, and they have to fight for the right to breed. They have male brains that enable them to carry out their duties as the troop demands, and as their assigned roles dictate. Females are likewise equipped for their role within the troop. Their bodies are adapted for that role, and so are their brains. They are the carers of the next generation who bear the young and rear them until they can become independent.

As a result, they have female brains to carry out their role. Male and female brains do not think the same because they are assigned to perform different roles in the troop or tribe. They do not think the same for this reason, and that difference between the sexes also gives cohesion to the troop and guarantees its survival. Male and female brains are driven by different goals and by their respective roles in each society.

The branch that broke from the chimpanzee family tree and evolved into humans would have behaved in a similar way, with the sexes following their own roles. Males were physically stronger than females, and modern males have inherited this aspect of their primate legacy. They had to be stronger for the same reasons male lions have to be stronger: to fulfil the roles that nature had assigned to them. The first human leader of the first human tribe had to fight for the top job. He had to fight to protect the tribe, and he had to fight to protect his harem. His body was equipped for that role, and so was his brain.

Modern man no longer has to fulfil this role because he transitioned from brawn to brain, but he still retains his muscles. They are a reminder of his past, and are still prized as a sign of strength. Strongmen competitions are held where men can compete with each other and male muscles are also attractive to the opposite sex. The competition for Mister World is not based on beauty, but on muscle size and physique. Body building is a throwback to an era when muscles were essential part of the male function in the tribe.

The women within the tribe were also equipped to fulfil their designated roles in the tribe. They had no need of strength because that asset was not required to perform the functions required of them. They were the bearers of the next generation, caring for their children and protecting them from danger. Their bodies had breasts to nurture the children with milk until the young could eat solid foods, the hallmark of mammalian society. The men hunted and gathered, and the women cooked the food and raised the next generation. The distinct and separate roles of the sexes was not defined by rules laid down by men or gods, but by the rules imposed by Mother Nature.

That is still how the world mainly lives, with men going out to work and women staying at home to rear the family. There are more working mothers now than ever before, but their role has not changed that much since the hunter-gatherer era. The bodies of men and women are obviously different and there is a good reason for that difference as explained. Men sire the next generation and women raise it, and that has not changed considerably over the millennia. There are same sex families now, but they cannot produce their own offspring and tend to go down the adoption route if they decide to rear them. The vast majority of both sexes obey the natural law by having their own children and rearing them into adulthood. The adults then leave home and complete the cycle by rearing their own children.

Male and female brains are different because of their respective roles in the tribe. Each had to fulfil the role assigned to it under the primate system of living together and sharing the same territory. Male brains were programmed to protect the troop in combat against rivals, and women's brains were programmed to rear and feed the next generation. Men and women did not think the same at the beginning of the human journey and they do not think the same now. Whilst there are many roles that men and women perform equally, the primate system still applies to the majority of people, no matter what part of the planet they inhabit or under what system of rule.

Regardless of colour or creed or nationality, women are the carers and the sex that rears the next generation, and men are the bread-winners. Changes in lifestyles and modern living do not alter that basic and universal rule. Men's brains have been designed for the role nature has assigned to them, and the brains of women have also been designed for that purpose.

Although we now exist in a politically correct age, proofs can seldom be disputed except by fanatics and these should now be forwarded to prove male and female brains are different. Boys and girls have different brains from birth. Their gender decides the roles they must perform in the tribe. Children are natural in their play and in their behaviour, until they learn how to live in the arcane world of adults and are expected to conform to convention and political correctness. Their future roles as adults have already been decreed by Mother Nature, and they can be observed in the way children play. Girls play with dolls and boys play with guns. They are acting out the future roles that has been assigned to them since birth.

Modern children are no different to the countless generations who have lived before, nor cannot be since nature determines their behaviour. Feminists and the politically correct brigades might wish for boys to be given dolls and guns to be given to girls, but that would be contrary to the laws of the natural world. The children would object too since their roles have already been assigned by Mother Nature for their future roles in society, and they cannot do otherwise but obey. Girls play with dolls because they are preparing to raise the next generation, and boys play with guns because they are preparing to defend their territory. And no amount of political correctness or wishful thinking can upset the laws of Mother Nature.

The transition from brawn to brain undertook by the distant ancestors of man rendered the former of lesser importance than the latter. The primitive ancestor who sought the alpha male position now had to use his brain more than his muscles. He had to figure out who he could trust and who would stab him in the back, literally. He had to figure out who would back him up if it came to a fight for the top job. He had to assess who could be bribed to betray a rival, or who could be induced to switch sides. When should he make his move? What were the consequences of failure? What would happen to him if he failed? Would be thrown out of the tribe? Worse yet, would he lose his life?

These considerations required a lot of brain power, and further stimulated the evolution of the male brain. Since females were not involved in the competition for the alpha male job, their brains went down a different pathway. They were not required to use them in the pursuit of power because their role in the tribe excluded them from the alpha male position. Also as already stated, their role in the tribe was already ordained, and it was not competing for the top job.

That the male brain consistently creates more geniuses than the female brain is not the same as stating one is superior to the other. Rather, it is to state that they went down different pathways because of their assigned roles in the tribe. Female and male brains are equal but different, and that distinction must be emphasized. It can be argued that women have been traditionally discriminated against by men, and they have been. It might also be argued that their role as mothers inhibited their ability to keep up with men in the intellectual field because they were more concerned with rearing their children than in the pursuit of knowledge. This proposition is true because their brains were designed to fulfil the roles that Mother Nature had mapped out for them.

The priority of the female brain lies in this task and not in the exploration of the mind. The male brain on the other hand did have to engage in mind exploration because it helped in the struggle to acquire the top job. Brain power was now an essential requirement for rule, and that was the most important asset in males. It still is for the man who wants to become alpha male or leader of his country. He has to figure out friend from foe, and who he can trust. He has to read faces and determine if the smile from a member of the gang is genuine or not. And that is within his own party, apart from the opposition.

The twentieth century saw the full emancipation and liberation of women in the West and the traditional roles were no longer fixed in stone. Women became aviators and surgeons, and could do their own thing without recrimination or censure. That century freed their minds fully for the first time in human history. Yet they did not create geniuses in the same quantity as men during that century. The Nobel Prizes were still dominated by men, particularly in the field of physics. The research into the nature of atoms, which led to the atomic bomb, was almost exclusively a male preserve. Male dominance in this field is no coincidence. The conception and construction of weapons has been part of the male brain since the first man constructed a club to defend his territory.

The importance of the race to develop the first atomic bomb has been largely forgotten, although its invention led to the hydrogen bomb and the capacity for mankind to destroy the planet and render it virtually uninhabitable. The fourth natural law of competition demanded that if one nation acquired an advantage, rival countries were certain to follow or face the consequences. No single country could be allowed to gain an advantage over a rival without a similar or superior response. The Nazis were involved in the race, and there can be little doubt that they would have used the device indiscriminately had they acquired it.

The Americans, aided by European scientists, won the race and dropped two atomic bombs on Japan to end the war. Then the Soviet Union acquired the bomb, which spurred a hotter arms race to develop more powerful weapons. Now Britain and France were compelled to join the race to catch up or risk being left behind. And when India built the bomb, its traditional enemy Pakistan had to follow or suffer a disadvantage. It can be seen from these examples that an advantage gained by one side must be counteracted by the other side under the fourth law. That is an immutable law of nature which must be obeyed.

Another field where men excel and women are almost totally absent is that of philosophy. This science concerns the search for truth and the meaning of life; but more crucially it also explores systems of rule. Confucius in China and Plato in Greece explored the most important question in the history of the world that is as relevant now as it was in their eras. Both philosophers discussed the system of rule most beneficial to mankind. Different types of rule are not essential to the female brain because of their assigned role within the tribe, but they are essential to the male brain. As a result, the female brain did not have to contemplate on the question of rule or the best form of government

Systems of rule were essential to the male brain because of their role within the tribe, and the possibility of getting that top job. Each brain concerned itself with the respective role that nature had assigned to it. Just as the body was designed for its assignment and function within the tribe, so too was the brain. Since the male brain was conditioned to explore systems of rule, it contemplated on that topic. And because the female brain was excluded from rule under the primate system, it did not have to contemplate on systems of rule. Male and female brains behaved in accordance to their designated role in the troop. Modern humans have inherited this legacy from the past, and the tandem evolution of male and female brains. Another example of these two different pathways can be found in weaponry.

The conception, design, and manufacture of weapons is overwhelmingly carried out by men because that is part of their role in primate society. Weapons take shape in the brain before they are turned into tangible objects of destruction, and the male brain was programmed to fulfil this task. The female brain took a different course, and the evidence of this difference can still be found in the modern world. The majority of hospital nurses are women, and the majority of army generals are men. The tandem evolutions of brain and body that occurred in both sexes. rains have assisted humankind to journey from the primitive world of the past to the sophisticated world of the present. Yet in that long journey, male and female brains have remained different because of their assigned roles.

When a woman acts like men in a role that nature has not assigned to her, such as the pursuit of power, she is invariably castigated and considered abnormal.

Pursuit of power is by a woman is regarded as unnatural because primate conditioning demands a man in the alpha male position. Society accepts, and secretly admires, men who kill rivals and scheme their way to the top through rivers of blood or devious cunning because these actions are part of the natural order. Should a woman use the same methods, she is roundly criticized for venturing into a man's world. Double standards of course, but a further example of the difference in roles that men and women are expected to play in the pursuit of power. A man can wade through rivers of blood to get to the top and earn respect for his dark deeds, but a woman who tries the same is castigated and condemned.

The double standards for men and women can be observed in *Macbeth* by Shakespeare. The ambitious wife of the title character encourages her husband to slay the rightful king and seize the crown for himself. Lady Macbeth is the bard's most notorious female villain, and she illustrates the difference between men seeking power and women seeking power. It's natural and acceptable for a man to kill for the crown, but it's unnatural and unacceptable for a woman to even urge him to commit his murderous deed. The primate rulebook decrees that power is a game only men can play, and women should have no active part in the proceedings. The villainous Lady Macbeth is notorious because she overstepped her role and ventured into a man's world.

The mantra is worth repeating that modern men and women do not think the same. Men are more organized and women are more intuitive, and these conditions did not come about by chance or by a roll of the dice. Men and women did not evolve to compete against each other in a battle of the sexes as happens in the polarized societies of the modern era, instead the sexes evolved in tandem to complement each other. They are programmed by Mother Nature to form partnerships and bring children into the world, in accordance with the third natural law. Men and women cannot be the same since they have evolved to fill different niches in the troop or tribe. Their bodies perform that role and so do their brains. The fact that they men and women think differently makes pair-formation work so well.

It would not work if their brains were the same.

PERILS OF PARENTHOOD

The demand of the third law that only the best genes must be passed down can be observed in the ritualistic contests for breeding rights. Males in many species have to prove their fitness to sire the next generation by trials of strength that are conducted to weed out weak genes. Females do not breed with the loser, but with the winner who has proved he has good genes and fit enough to produce healthy offspring. Sealions, deer, mountain sheep, goats, and males of many other species compete in battles for access to females, and some are killed in the contests. Others are rendered too weak to escape predators and provide easy victims.

The fights are designed by nature to expose their weakness and remove them from the breeding cycle, thereby ensuring that only the best genes are passed down. Whether by fighting or feats of endurance, the weak genes are removed and the strong genes promoted and passed down. Breeding the next generation is not a right, but a demand imposed by the third law in trials of strength. The battles must be fought and they must be won to guarantee that none but the best genes deserve to survive into the next generation.

There is a minority of the human population that apparently does not obey the third law and that segment should now be looked at. It is not the purpose of this work to determine the causes of homosexuality, but rather to demonstrate how the natural laws are universal regardless of sexual inclination or orientation. There are no reliable figures for the size of the homosexual community, either because of privacy laws or due to the stigma it carries in some societies. Estimates hover around ten per cent of the population, but this figure is of course unreliable. Some undoubtedly choose to remain in the closet where it's safer. To examine how the homosexual community conforms to the immutable laws of nature, a theory is required. Unfortunately, no facts are readily available to back it up and prove it as truth. Instead, the available evidence shall be presented to give the theory credibility.

This theory suggests that this community channels the third natural law into creative art because the urge to create is part of nature. The third law of nature is not arbitrary, since every species is born to propagate itself and pass down its successful genes. As a general rule, homosexuals do not pass on their successful genes by having children of their own. Modern homosexual couples are also driven by the third law. They can, and do, adopt children but that is not the same as perpetuating their own genes as the law demands. The genetic engine that drives life requires a continuation of genes that this community fails to fulfil. Another area must therefore be explored to discover how they obey the third law, and that can be found in art.

Bringing children into the world involves an act of creation, and producing art also involves an act of creation. Both acts are driven by a desire to leave something behind in obedience to the third law. The joy that parents feel on the birth of a child is the reward by nature that the third law has been fulfilled. The joy that the artist feels on completing his work is a similar reward for the creative process. Homosexuals far outnumber heterosexuals per head of population in the field of art, and there must be a reason why they are more prominent. The caring parent who creates a child is concerned with its welfare and devotes his whole life to rearing the child until it reaches adulthood. The artist is concerned only with his art because that is his creative offspring. Art can last as long as a family tree, and in the case of great art it can outlast most family trees. Some examples are now used to demonstrate how homosexual artists channelled their creative urge into art, and in the process obeyed the third law.

The greatest artist and creative genius in the history of the world was homosexual. Leonardo da Vinci lived in a society where his sexual proclivity demanded the death sentence. In fact, he was accused of sodomy as a young man but the charge was not proven in court, otherwise the world would have lost its greatest artist. The world without masterpieces such as *The Mona Lisa* or *Lady with an Ermine* is difficult to imagine. His sketches include designs for a tank, a helicopter, and underwater breathing apparatus. His drawings on the human body, which he obtained by dissecting corpses, are astonishingly accurate in their detail.

A contemporary of his was Michelangelo, the supreme sculptor in the history of art and another homosexual. He painted the roof of the Sistine Chapel crouching on his back that still draws millions annually to Vatican City. His work is astonishing, and creates a hushed silence on the throngs who gaze up from below, and instils a sense that they are in the presence of the divine. He created the statue *David* from a single block of marble, a statue that remains the pinnacle of sculpture. It gives the appearance that the artist had released life itself from stone. He too was obliged to conceal his homosexuality because of the death threat hanging over his head.

Anti-homosexual religious laws, coupled with the real threat of death if discovered, combined to keep many artists from revealing their true sexuality. Some were forced to conceal their homosexuality just to stay alive, and earn a living by selling their art to rich patrons who also offered protection. The Medici family in Florence, who were the godfathers of the Renaissance, not only offered commissions to the artists in the city but also their protection.

Two of the greatest painters of the twentieth century were Francis Bacon whose work commands millions, and Grant Wood who painted *American*

Gothic. Both were homosexual. The list of gay writers reads like an encyclopaedia to literature, starting with Sappho, a Greek poet who lived on the island of Lesbos from which the word *lesbian* is derived. More modern names are also worth mentioning to illustrate the prevalence of homosexuals in art. Oscar Wilde the playwright, Walt Whitman the poet, Virginia Woolf the writer, Truman Capote the writer, Tennessee Williams the playwright, Marcel Proust the writer, and T.S. Elliot the poet. These are but a few of many whose names are recognizable.

The stigma that was once attached to homosexuals in music has largely disappeared, and now they are out in the open. They are far too numerous to mention here, but it was not until relatively recently that they could openly admit to their sexuality because it carried a prison sentence in some Western countries and the death sentence in theocracies. Composers from the world of classical music were forced to conceal their homosexuality because of the repercussions they would face. Franz Schubert and George Frideric Handel were both homosexual composers. Russia's greatest composer, Tchaikovsky, who penned *Swan Lake* and *The Nutcracker* also belonged to the minority segment of humanity. No doubt there were others who concealed their homosexuality because of prevailing taboos and dangers, and were forced to live out their lives in a lie.

The majority has always persecuted the minority who were different, and this is particularly true with the homosexual community. The umbrella of art, and the protection of their patrons, provided a safe refuge where they could channel their creative energy, and escape the censure and persecution of their fellow citizens. These artists did not perpetuate their own genes, but they did channel the creative urge inherent in all species into their art. They too obeyed the law, albeit through their great art.

Perhaps the theory has been proven.

SYSTEMS OF RULE

There is no more important question in the world than how are the people ruled. The excitement generated at elections testifies to their importance, and they are one of the few topics that can dominate the airwaves and viewing platforms, and render other programmes less important. People must know who shall lead them into the future and that is the reason why elections provoke such interest. As already stated, every system of rule is based on the primate template which requires a leader at the top and a gang of helpers. That basic model cannot change, but there are some variations and they shall be now examined and dissected.

For the majority of human civilization, mankind has lived under the rule of kings and emperors. These systems evolved from successful warlords who created dynasties to keep power in the family, and pass down their territorial gains to the next generation. It should be noted that hereditary rule excludes those outside the royal family from gaining power, unless by revolt or revolution to overthrow the regime. Royal families began to believe that blue blood ran in their body and that placed them above the common people who had red blood, and therefore they were entitled by birth to rule.

They ignored the dire warnings of nature about incest and married amongst family members because they had distanced themselves from the natural laws. They had come to believe that their blue blood would be tainted and diminished if they mixed it with red blood, and that is the reason European royal families intermarried, treating marriage like an exclusive closed shop. One of these alliances would end the rule of a dynasty and change world history forever.

More than two millennia ago, Republican Rome tried a novel idea to rule its growing empire and one that has since been emulated by the USA. No king could rule in that society because of its history, and they came up with a system of rule by three men. It was, in theory, a good idea and designed to check the ambition of a single leader. His two co-rulers were expected to move against him should he try to seize power for himself. It did not succeed in Rome because it broke the golden law of primate rule. Any system that contravenes the primate rulebook can never hope to succeed. There must be an alpha male at the top, a single person that the people can identify with and acknowledge as their protector. This primate legacy from the past is not open to change nor contradiction. There cannot be more than one leader nor a shared leadership for that basic reason. Any system of rule that deviates from this fundamental primate principle is doomed to failure and must collapse.

The same primate demand applies to this day. The people must know who shall protect them in their hour of need, the most important attribute in the leader. No Roman citizen knew who was the alpha male with three people at the top, and that was the system's fatal flaw. In these situations, one man always rises to the top whether through force of will or the strength of his army. In Rome, Julius Caesar became the alpha male and started a civil war that tore the Roman Republic apart and paved the way for the first emperor.

Thus far, it has worked in the USA because the system established three branches of government with equal powers. The checks and balances built into the constitution were specifically designed to prevent a king from taking control of the country. There is the Executive as represented by the president. There is the Legislature as represented by the congress, and there is the Judiciary as represented by the supreme court. The system has worked for two and a half centuries, not only because of the checks and balances, but more importantly because of the alpha male position as represented by the president. The people can identify with this figure, and he is expected to protect them in their hour of need. Not only that, the country's allies also must know who the leader is because they too rely on him or her for protection since he is the leader of the alpha male country.

The rule of kings in Europe also broke the natural law, but in another way through intermarriage between blood relatives. Princess Alexandria was a granddaughter of Queen Victoria, and she married Czar

Nicholas of Russia who was also related the queen. After four unsuccessful attempts for a male heir and producing only daughters, their fifth child was a boy they named Alexei. He carried a gene that does not allow the blood to clot and heal, causing the condition known as haemophilia. Any bruise or cut could have been fatal for the child and there was no known cure at the time.

The czar's wife turned in desperation to a mystic named Rasputin who seemed to have miraculous powers; but soon rumours were spread that they were having an affair, and ears are always willing to believe gossip when people are not happy with their lot. Events conspired against the royal family and riots broke out on the streets. The revolution that followed killed the czar and his family, and ended the dynasty by extinguishing the Romanov genes. Communism replaced three hundred years of royal rule in Russia, and the ideology was exported to stir up worldwide revolutions. Incest is contrary to the laws of Mother Nature, and there is always a price to pay for breaking one of her sacred taboos.

Like a spectre from the past, the military junta makes an occasional appearance on the world stage. The military takes over a country that is not being run to their liking. The junta is yet another variation of the primate system, with a general or a senior officer as the alpha male and aided by a gang of high officers in uniform. The problem is that men who wear uniforms think in the same way and are trained to obey instructions. As a result, the junta tends to treat the civilian population as soldiers. When orders are issued to the population, the junta expects civilians to obey as if they were soldiers. Imagine their surprise when civilians do not behave like soldiers and refuse to obey orders. This is the reason why military juntas have very short shelf-lives, and are always a temporary arrangement.

Theocracy, or rule of the priests, is undoubtedly the most backward system of rule to have ever evolved and arguably one of the most blood-thirsty. This is not to say that every theocracy is steeped in blood, only those with a single god or dominant thought that denies freedom to think outside official dogma. The religions that worship many gods, principally because of their diversity, are more tolerant and can accept new ideas. Monotheism is the religious equivalent of secular dictatorship as both are based on a single ideology that brooks no rival. Here is the basic cause of its inherent violence and why it has waged wars against those who are different. History once more provides an example of the religious dictatorship.

During the last century, the shah of Iran was toppled from power by a religious movement headed by Ayatollah Khomeini who had been granted asylum in France. He detested secularism and Western values, and constantly condemned them in his sermons. Religious leaders are not renowned for their sense of irony. Iran, formerly known as Persia, was a tolerant society and not given to extremes of religious fanaticism. That tolerance would soon change when the ayatollah seized power. He established a reign of terror to crush any opposition to his rule, including setting up a secret police force. No dictatorship worth its name can function properly without the secret police. He publicly hanged his enemies in batches of six at a time, including children as young as thirteen. Commuters on their way to work in the capital Tehran were daily reminded of the price for dissent by the corpses hanging from building cranes.

Every dictatorship behaves in the same way, whether religious or secular. The dictator rewards his enablers with top jobs and various perks, and the ayatollah rewarded the clerics by giving them high positions in his administration. Monotheism has always been about power in this world and not spirituality in the next. Religious rule in Iran followed the primate model, as it must, with an alpha male at the top aided by a gang of enablers. There have been several uprisings against the rule of the clerics, all of which have been violently crushed. However, rule of the clerics can have no future because it's exclusive rather than inclusive, and therein lies its fault lines.

A system that denies change at the top is contrary to primate system of rule that demands new blood to strengthen the gene pool. Religious rule is reinforced by imprisonment or death for those who fail to obey. The ideology of the leader is used to maintain his grip on power, and cement his followers to his side. They do not question his right or authority to rule because of the dominant thought he implants in their minds. But not all people think in the same way, and not all buy into the same religious ideology that the leader seeks to impose. As a result, the dictator must clamp down violently on dissidents and those who question his rule.

This is the reason why every dictatorship is intrinsically violent and why they lack basic human rights. Primate rule demands change at the top because change is part of the system that ensures the best genes are passed down. The alpha male must undergo challenges to establish his right to rule under the primate system. The dictatorship cannot change because the dictator seizes power for life. Change cannot evolve under this system of rule, and as a direct result can only be brought about violently, or by the death of the dictator. The dictatorship usually collapses internally when the people eventually rise up and end it in blood. The dictator always forgets that the real power does not rest in his hands, but on the streets with the people.

The evolution of democracy required a new system of rule that undermined the primate system that had remained unchanged since man evolved. It required disparate tribes with their own leaders to accept the rule of the leader of another and rival tribe by peaceful means, and share the same piece of territory. Before the evolution of this system, one tribe had to conquer its rival by force to have them live in the same territory and be governed by the conquering leader. What made the new system much more successful than any that had gone before was its flexibility and willingness to evolve peacefully.

Leaders of other clans or tribes were not excluded permanently from power, but could gain the highest job at the next election. There was now no need to remove the leader by violent means because he could be removed by a vote. This is the reason why democracy is the least violent system of rule that has ever evolved.

Although democracy is a departure from the primate system of rule, it does follow some aspects. It allows for change at the top at election times, a vital requirement under the natural law. Not only is there regular change at the top, the ruling ideology also changes. A right-wing party might win power or a left-wing party might win power, depending on the will of the people. The electorate accepts the changes because they have agreed to participate in the system. If they do not like the policies on offer, there is no need to change the leader by violent means. He or she can be dumped at the ballot box and bring about a new policy.

The people who believe in a socialist system can be accommodated as well as those who believe in a capitalist system. The transition from one system to the other is usually seamless and peaceful, at least in long-established ones. The system is also more dynamic because it brings change in its wake with each new change of government. No society can advance without diversity at the top, and that is an essential requirement of evolution. Change at the top is also required under the primate system of rule, and here is the reason why democracies are usually stable. As primates, people expect a challenge for the alpha male position and become restless when it fails to happen. This is reason why dictatorships are usually unstable and seethe with discontent beneath the surface.

Democracy is based on trust and on the willingness of the defeated leader to accept the will of the people. That requires him to accept defeat gracefully and step down, and pass the reins of power to the person the people have chosen to lead them. Should he question the legality of the election, or claim it was rigged, violence is a natural consequence. There is no more seductive mistress than power, and no leader wants to give it up, but the system demands that he does. The people are the real power, and history has proven that old adage time and time again. When they speak at the ballot box, he must listen or a civil war could follow. Trust is a double-edged sword, after all.

The transition from brawn to brain that occurred only in mankind and in no other primates enabled the evolution of democracy, and also cleared the way for a woman to get the top job. This fundamental change was a complete rupture with the primate rulebook that demands a male at the top. When the first democratic elections were held in Athens more than two thousand years ago, women were not eligible to vote. It has taken almost as long for them to have equal voting rights with men in democratic elections, and they had to fight for their rights. Now they can become leader of a country and no eyebrows are raised.

This is yet another example of how thought evolves to meet the demands of a modern society. Democracy does not impose mental chains on the mind like the theocracy or dictatorship, but permits it to move forward and evolve to meet changing times. It also allows for a change at the top, a vital demand under the primate system of rule. As thought continues to evolve, politics will likewise evolve to meet the changed world. Secular dictatorships shall always be here because of primate evolution; and the dictator is usually seen as a strong man who can fulfil the role of protector, the primary function of the leader.

The religious dictator has no future because he is sectarian and cannot regard competing religions as equals. Therefore, he must crack down on them, creating a sense of grievance and division within his own society. That is precisely what happened in Europe under the religious dictatorship of the popes, when dissidents and opponents to the rule were hunted down and killed. One of the worst atrocities in that period of religious dictatorship was carried out in southern France on a peaceful community known as the Cathars. This was a peaceful sect that lived by charitable deeds and healing the sick. Their problem was not that they were more Christian than the pope, but that did not buy into the ideology that the pope sought to impose. They had a religion of their own and lived by its teachings. In response, the pope organized a crusade that wiped out the peaceful community, and thereby sent out a warning that the price of disobedience was genocide.

The system under greatest threat is the democracy, and there are reasons why it must change or drown in the rising tide of expectation. Democracy is in danger because it is failing to obey the primate rulebook. True, it has abolished the need for violent change, but it has forgotten the golden rule in the primate book of leadership. The people must know who is in charge because that knowledge is fundamental to their state of mind, and more especially to their sense of security.

The modern democratic leader is widely regarded as a bureaucrat and not as an alpha male. The civil service is now seen as pulling the strings of power behind the scenes, and that speculation always leads to conspiracy theories. Instead of listening to the people who elected him, the modern leader abandons them after election and listens to his civil servants and spin doctors and marketeers who remain hidden. The people do not see an alpha male leader, but a puppet moved by invisible strings.

The European Union as it currently exists cannot last for long because it ignores every page in the primate rulebook of power, and the system that does cannot hope to survive. There are certain fundamentals required for any system of government to rule, and the most important one is choosing the leader. The EU is a community of nations, all of them democratic. Yet the president is not democratically elected by the European people, but chosen inhouse like a private golf club open only to a select few who dictate the rules. The candidate for the alpha male position is recommended for the job by the European Commission. The European Parliament then votes, and either accepts or rejects the candidate. By excluding the people from the most important decision in every democracy, which is electing its leader, the elites have excluded them from participating in the process. The EU therefore cannot credibly claim that the system is fully democratic. Instead, the EU is an elitist club.

What the people in every society must know is who shall protect them in their hour of need. The selection of the president, as occurs in the EU, is not conducive to filling them with confidence and hope in the future. The lack of strong leadership at the top became apparent during the refugee crisis when a million people were seeking refuge on the continent from the war in Syria. Not all were genuine refugees entitled to sanctuary, but the EU was lacking in leadership and behaving rather like a headless chicken in search of its lost head. The flood continued unabated, and rumblings of discontent were voiced in Greece and Italy who were on the front line. Instead of sorting out the problem, the EU behaved in its typically bureaucratic manner by washing its hands of the problem and leaving it up to someone else.

It effectively reneged on its duty of care to its citizens by ceding control of its southern border to a country that's not even in the club. The EU bureaucrats bribed Erdogan of Turkey to turn off the tap of refugees and perform the task it was supposed to do, defend its borders. Defence of borders is an integral part of the primate condition, and the EU failed miserably to live up to its obligations and its duty of care to its citizens. By bribing Erdogan, it has given this would-be dictator a bazooka that he can hold to its head whenever he chooses. He can turn on the flow of refugees like a tap when he feels like it and demand more money. Blackmailers have a tendency to always come back for more. The disillusionment with the failed response to the refugee crisis, added to the lack of leadership at the top, has forced countries within the bloc to have second thoughts. Who wants to live in a system that lacks leadership and direction? Not every system of rule is immune from the voices in the streets, and bureaucratic rule is no different.

The EU has consistently ignored the primate rulebook and that is the reason it's coming apart at the seams. The strong leader is beginning to emerge again because the people require that innate sense of protection. Poland and Hungary have elected strong leaders who are seen to protect the people. More are certain to follow because the vacuum at the top of the EU is unsustainable. History has shown time and again that when people are afraid and their fears are not addressed, then real and permanent change is brought about violently.

The more things change, the more they remain the same; and the future belongs to the secular dictator. He is the man with a clear purpose and the vision of mind to tell the people what they want to know. More importantly, he tells them that they are safe, that he will protect them from both external and internal enemies. He proves his strength by killing his enemies, and the people love him for getting rid of traitors and dissidents who work for foreign governments. The more enemies he kills, the safer the people feel. He is the latest incarnation of the tribal leader who protected the troop and stamped out dissent by either killing or exiling his enemies.

The secular dictator of the future shall have many willing enablers to assist him to achieve his goal, and to maintain him in permanent power. The primate system, after all, is based on this basic model, with an alpha male at the top and a gang of enablers. There are many reasons why the dictator never has a shortage of enablers when he seizes power, and who are eager to follow out his orders without question. Some of these are worth reiterating to prove that dictatorship is here to stay. It is the oldest form of rule and has not lost its alluring appeal because it adheres strictly to the primate model. The leader's enablers are programmed by nature to put their trust in the alpha male, and to obey him whether he is right or wrong because that was how the first human troop operated. Rule has nothing to do with morality or choosing between right and wrong, Rule has nothing to do with good versus evil either, as still can be observed in present dictatorships. Rule is solely concerned with gaining power, and then staying in power.

The enablers are also motivated by the first natural law, which is survival. The people who help the dictator have a far better chance of keeping their heads than the people who oppose him. Self-preservation is the most compelling argument for supporting the alpha male in power. Being close to the dictator also grants the enablers opportunities to influence him, and to make their enemies his enemies. They can get rid of their enemies by having a quiet word in the ear of their master. And the people respect the enablers because they can also intercede with the dictator on their behalf. An enabler can have a word in the ear of the dictator to save the supplicant from imprisonment or death.

Helping the dictator is also great for the bank balance, and money is a powerful incentive that often trumps nationalism, or a sense of duty to the country. Enablers get wealthy by propping up the dictator and keeping him in power. They get their share of looted property or any goodies on offer, just as the helpers who aid the chimpanzee alpha male get the choice pieces when a monkey is killed for food. The enablers can acquire oilfields appropriated from foreigners, or mining rights for the country's assets like gold and silver. They can strongarm private businesses and set up protection rackets in the knowledge that they are immune from the police. They can launder their dirty money abroad, and find banks that are always willing to turn a blind eye to their crimes. Banking too is not about morals or distinguishing right from wrong: banking is about one thing only, and that is making money.

Whilst the future must remain uncertain, the basic system of rule is not open to change. Whether that future lies in the hands of the dictator or in the hands of the democratic leader, the basic model must remain the same because it has been ordained by Mother Nature. There must be one alpha male at the top, be it a man or a woman, aided by a gang of enablers. That leader must instil in the people the vital sense of security for their psychological welfare and peace of mind. They must have watertight assurances that they shall be protected in their hour of peril. Their home territory must have secure borders that are clearly marked out, and trespassers punished and seen to be punished severely. Any aggressor who seeks to invade and seize the territory must be warned that the cost is too high.

Rule hasn't changed all that much down the ages.

JOB DESCRIPTION OF THE ALPHA MALE

To demonstrate the subtle manner by which Mother Nature profoundly and continuously influences mankind, comparisons are now introduced to prove her subliminal control of the human mind. A man or woman who desires to become an airline pilot does not immediately sit into the cockpit and try to take off the runway. He or she would not be allowed to fly that plane without proper training which takes many years to perfect. And if that person tried to fly the plane without proper training, the result would end inevitably in a crash. The same basic lesson also applies in the field of medicine. A man or woman who desires to be a surgeon would not be permitted to enter an operating theatre and perform surgery on a patient without the required and proper training. That person would not have a clue how to safely operate simply because he or she would not know what they were doing. The required knowledge for the job is not stored in their memory banks because it has not been implanted by training.

Now, compare these two examples to the alpha male leader in prehistory who achieved his position by fighting for that top job and cultivating a gang of enablers to aid his rule. He knew how to do his job because Mother Nature had implanted the necessary and required knowledge in his memory banks to perform that role. He was not stepping blindly into the unknown, but was fully equipped for the task in hand because he had the correct tools in his head. That implanted subliminal knowledge instructed him to establish and mark out the borders of his territory with visual signs of ownership, and reinforced by auditory warnings. He knew the course of action to take if his borders were transgressed by a rival tribe who sought to take it over. He was in the forefront of his tribe when reacting to that threat and seeing it off by killing the intruders or chasing them out. If a challenger arose from the ranks and attempted to topple him, he knew how to meet the challenge by tackling it head on, and calling on his enablers if the going got too tough.

One of the most important requirements for the job is a demonstration of fitness to rule, which sends out the message that the tribe is in safe hands. He demonstrated his fitness to rule by quelling internal disputes that might damage the cohesion of the tribe, protecting vulnerable members from aggression by upstarts, and dishing out beatings to demonstrate he was in charge. The beatings served to emphasize his strength and therefore his right to lead the tribe. A male member of his tribe who approached without showing due deference to his authority and position was chastised. That alpha male leader did not need lessons on how to react, he had been programmed by Mother Nature for these precise situations. He knew too that his rule would be challenged from time to time because he lived within the immutable laws of nature that demands change at the top. He knew too that one day he would be toppled by a younger and a stronger male according to the laws of primate life and society, but he would cling onto power for as long as his enablers backed him up. They were his power base that he relied on for support. When the inevitable day came and his enablers failed to come to his aid, he undoubtedly also new that his limited term as alpha male was finished and power had been conferred on a stronger rival.

The modern alpha male leader has inherited many aspects of that counterpart who lived tens of thousands of years ago in an unknown forest in an unknown region of the world, because he too is a primate and subject to the rules of that society. The alpha male is always challenged because that is a natural primate law which demands change at the top to introduce new genes, and every leader who takes up the reins of power know this. That natural law is imprinted on his memory banks. The modern dictator can have his challengers murdered, locked up on false charges, or disappeared in the night, as long as his enablers back him up. When they see the writing on the wall and begin to desert him, he too knows that the game is up. This scenario is not new, but has repeated itself again and again in pursuit of the alpha male prize. There is no greater reward in life than ultimate power, and there is no greater loss in life than ceding that power to a stronger and fitter rival.

The requirement to display fitness to rule can be observed in modern leaders, a legacy from the past that is subliminally imprinted on their mind. The strongman is seen across the world presiding over a display of his country's military might on May Day or on Victory Day, or on his birthday. A strong military is the alpha male equivalent of a strong leader. That is his projection of power politics, which sends out a message to his enemies that his country is strong, but more importantly that he has the strength to crush an internal revolt. The enemy within is more dangerous to every dictator than the external enemy, and military parades are a reminder that he can use the army to enforce his rule if they try to topple him. This is what happened in Germany when Hitler compelled the army to take a personal oath of loyalty to him, not to the country. The army was his tool of enforcement after that oath of allegiance, to be used by the dictator against his enemies.

Leaders in a democracy too must display their fitness to rule too because that is part of the CV as laid down by Mother Nature. It is a requirement for the job that they do not know exists, but must comply with because they cannot disobey the alpha male blueprint. Golf would appear the sport of choice for Western democratic leaders to display their ability to rule and that sport would appear to be a prerequisite for American presidents who spend almost as much time on the golf course as they do in the oval office. Golf is portrayed as a noble sport without cheating, and this aspect reflects on the leader and on his country. Golf is also a non-violent sport which also sends out the message that the leader does not believe in violence. By walking around a golf course, the leader exhibits his fitness to rule. And by marking the score card, the leader displays his mental acuity to lead. If he can perform these tasks successfully, the leader is fit enough to rule the country.

When Margaret Thatcher secured the top job to become the first female prime minister of Great Britain, she too had to display her fitness to rule as required for the alpha male position. As a woman, Thatcher had to prove that she could perform the traditional male role that had heretofore excluded females. The change in her was immediate, most especially in her voice. It became deeper and less feminine, and began to sound almost masculine, as if emulating the voice of a male leader. She was also seen in a battle tank, sending out a visible message that the country was in good hands, and that she was capable of defending it against its enemies.

Women who stepped into the top role had not only to overcome male objections, they had to overturn three hundred and fifty thousand years of primate conditioning that demands a male as leader. They had to display their fitness to rule by reassuring the people that the country was safe under their leadership. Catherine the Great of Russia can be used as an example to prove this point. She elbowed her neurotic and incompetent husband aside to become Empress of Russia, one of the great European powers. As a woman in a traditional male role, she had to demonstrate to the people her qualifications to protect them from attack, the most important consideration for the top job. She had a portrait painted depicting her in a magnificent military uniform riding a cavalry horse, and holding a sword aloft in a threatening pose. By adopting male trappings of power and prestige, she sent out the message that the people were in safe hands.

A similar message was also sent out by Queen Hatshepsut of Egypt who became pharaoh. She too adopted the male trappings of rule and power by wearing a false beard and the regalia of office. She was depicted on monuments smiting the enemies of Egypt in the same way male pharaohs were shown. These were visual displays of her fitness to rule, and sent out a powerful message that the people under her protection were in good hands. The visual displays reassured the people that they were safe under her rule, and that she was quite capable of protecting them from attack. Protection is the essential requirement for the leader in primate society.

The above examples prove that alpha male leader, be that male or female, does not need a manual for the job, nor a course on the rules and requirements. Mother Nature has already imprinted the duties and the responsibilities demanded for the alpha male role. And since women have been excluded from that role for many millennia, they must adopt male trappings and mannerisms to demonstrate a fitness to rule. They cannot unilaterally change the rules because they are imprinted on the mind and no force on the planet can remove them. Territory must be secured and defended against allcomers, both internal and external. The loss of territory can literally result in death, and that is why wars are so violent and bloody since all wars are fought for control of territory. Civil wars are more violent and vicious than most because both sides are fighting over the same territory, over the same strip of land that supports them both. Each side may claim that the war is fought for ideological purposes. They may claim it is socialism versus capitalism, or religion versus secularism, but they are fighting for the land itself, for territory.

Land is the ultimate reward, and here is the reason why peace treaties have invariably granted territory to the victors. After the war is won, the victors sit down to carve out their slice of territory from the conquered foe and claim it as their own. Wars can be divided into two distinct and separate classifications: wars of aggression fought for more territory, and wars of defence fought to preserve or recover lost national territory. A sense of nationalism is not required to defend home territory since that duty is hardwired into the human psyche. That duty to defend home turf has existed before the nation state evolved, and indeed before a sense of nationalism came into existence. Defence of home turf is common to all territorial animals no matter where they live because that message cannot be disobeyed.

Not a single year has passed without a war in the recorded history of mankind. That sentence is worth repeating, not a single year has passed without a war in recorded history. It can also be stated, without fear of contradiction, that not a single year has passed in the evolutionary history of mankind without one. War is not diplomacy by other means as sometimes portrayed, nor to right ancient or perceived wrongs, but rather the extension of territory by violent means. That is the sole purpose of war, which is part of the natural order for territorial animals under the second law of nature. If territory is not defended adequately and ferociously, it will be taken over by a stronger rival.

Living in the twenty-first century does not change the second law of nature as recent events in the Ukraine prove. The seizure by Russia of Crimea, although seen in the West as a land grab, was seen in Russia as the recovery of national territory. The Crimea has traditionally been regarded as Russian national territory that was ceded to Ukraine without the consent of the people. There can be no greater loss of face for an alpha male than the loss of territory because that renders him weak in the eyes of the tribe. No leader who is seen as weak can hope to remain at the top for long, and shall be deposed since he has shown weakness. Putin bolstered his image in Russia in the eyes of the Russian people by recovering that lost territory. He was the strong leader who would offer the Russian people protection after the debacle of Soviet Union breakup, and he still trades on that reputation to maintain power, not to mention silencing his enemies. His wholesale looting of the country for his personal gain can be overlooked by the same people because he is regarded as a strong leader who has restored lost territory.

The third world war will probably erupt in the South China Sea where tensions are already high. China is grabbing isolated islands disputed by its neighbours in this competitive sea that is an essential trade route. It has filled in tiny coral reefs and turned them into military bases complete with missile installations, and then claimed territorial waters rich in fish and natural resources like oil and gas. It is also disputing islands that Japan claims as its own and is prepared to defend; and using threatening language against the island of Taiwan. It regards this island as a lost province that must be recovered. Taiwan has a very close relationship with the USA, and regards that country as the alpha male who will come to its aid if China attacks. Wars are always unpredictable, but there can be no winner if it turns nuclear.

China is already large enough and does not require more territory. It cannot afford disputes with its neighbours that could escalate into armed conflict, damaging its economy perhaps irretrievably. The country is far from united with an independence movement in the annexed province of Tibet, and a large Muslim majority in the province of Xinjiang that seeks independence and its own homeland. The fight for democracy in Hong Kong hasn't gone away either, and China could split into warring regions in the event of an armed conflict with the USA and its allies, who would certainly exploit the country's internal divisions. But that is not the message its alpha male leader hears because the need for more and more territory is part of the human condition.

There is another message playing out that is part of primate society, and one that the likely protagonists are unaware exists because modern man has become separated from his origins. Mother Nature is dictating the unfolding events in the South China Sea, not the USA and not China. After years of the Trump administration that exposed the internal fractures in American society, the USA is now viewed as a weak and fading superpower. The withdrawal from the world under the Trump doctrine has left an alpha male vacuum to lead the world, and primates detest a vacuum. There is a parallel to be found in the chimpanzee rulebook. When the alpha male is regarded as weak and losing his power, the number one challenger starts throwing his weight around to provoke a reaction. He is spoiling for a fight to challenge the alpha male's fitness to rule. He ramps up the pressure by shows of strength and attacking the enablers of the alpha male, who must react to the increasing threats to his authority or lose face and his enablers. The USA is the current world alpha leader who is considered weak, and China is the number one challenger for that role. Hence the provocations in the South China Sea, and the proofs that mankind is not in control of its destiny.

No country nor society is immune from the second law, and its neighbours are also likewise governed by this same law which is universal to all territorial animals. They cannot lose face by not responding to the threat because that would show them up as weak, an unforgivable sin in the alpha male that the populace will not tolerate nor forgive. The job description of the alpha male requires him or her to defend national territory, and no leader is likely to stand idly by and let China do a land grab on territory he or she regards as home turf. The response to China's moves was inevitable and predictable, of course. Taiwan is beefing up its defences and buying more weapons from the USA. Japan is already heavily rearming despite its peaceful constitution that was introduced after its defeat in World War Two. The Philippines too is buying anti-ship missiles to deter China's threatening moves against its outlying islands. The fourth law of nature is playing out in that region, as it has played out for countless years. An advantage by one side in military hardware and weapons systems must be met by an equal or superior response, and this fourth natural law cannot be disregarded.

The primate siren calls for the defence of national territory is far more potent than any written document drawn up by man, and China is certain to meet stiff resistance if it seeks a landgrab. The message is not written on fragile paper that can be ignored or torn up, but indelibly imprinted on the brain where it cannot be disobeyed. Neither can it be removed, nor indeed altered. Man is a territorial animal who is prepared to put his life on the line in its defence. There are also other countries in the region with legitimate territorial claims in the South China sea and they too are preparing for armed conflict. This same war dance has been played out for ages, this same response to a hostile armed threat against national territory.

A wise Chinese alpha male should consider the future before starting a conflict that could destroy its economy and country, but such consideration requires clear thinking. The importance of territory is deeply ingrained in the human psyche, making it unthinkable to think clearly or rationally. In a conflict where there can be no winner if it turns nuclear, it would be more rational to seek a peaceful solution in dialogue. However, that is to ignore primate conditioning which is not achievable because the need for more territory cannot be ignored nor countered. That same need has been the underlying cause of every war in the history of the world, and the reason why empires were created in every inhabited continent on the planet. They were forged in Africa for more territory. They were forged in

America for more territory. They were forged in Australia for more territory. They were forged in Europe and in Asia for more territory.

Territory is always worth fighting for, and territory is always worth dying for. History teaches that lesson time and again, except its lessons are never heeded.

Mankind praises peace although he seldom practises it, and offers the Nobel Peace Prize to those who end conflicts and disputes. Peace prizes suggest that man is in control of war, whereas Mother Nature is in control of war. She calls the shots and always has, and that pun is quite deliberate and relevant. Territorial animals wage war for more territory, mostly with claws and teeth and strength of body. The territory must be defended against a takeover or it will be lost. Man is better at it than any other animal that has ever walked on Earth because of his advanced brain that allowed him to make terrible weapons of mass annihilation.

He now has the ability to wipe out every living creature on the planet. The fact seems to have escaped him that nuclear weapons would destroy the very territory he is fighting for because radiation fallout would render it useless for crops or livestock, resulting in famine on an unprecedented scale and the loss of countless lives. A nuclear desert could not support human life, but the need for more territory is so hardwired in the human brain that it prevents rational thought or reasoned debate. That overriding consideration relegates all other thoughts and considerations, and renders them irrelevant.

The alpha leader who leads his tribe against a neighbour is always revered by his followers. He is greeted as a hero at home and welcomed with parades and honours. It matters not to the welcoming crowds that their hero has killed fellow human beings and stolen their land. The cheering crowds hear the subliminal message that more territory guarantees the survival of the tribe. Extra territory can also support a larger tribe, and there is strength in numbers. But every coin has two sides, and sometimes it does not land the right way up. The downside is that the tribe will desert him should he fail to win more land and brings the enemy to their gates. He can expect no forgiveness from the people he led into ruination, and sometimes no mercy.

The great empires that rose in the morning sun and fell in the darkness of night were forged for more and more territory. The alpha males who forged them have gone down in history as military geniuses and their famous names resonate to this day. They are seldom criticized for the deaths of millions, nor the theft of land and property. Every war since man first took up a club to invade the territory of his neighbour has offered ripe opportunities for theft, except during a conflict it is called booty. That word is a nicer term on the ears than downright theft, and negates any sense of conscience the thief might feel over his actions. Theft is a sour crime that leaves a bitter taste, but booty is the sweet wine of victory.

The great conquerors are widely admired because of the vital importance of territory to every society and people, and debates still rage as to which empire was the largest in history. The peoples whose lands were stolen and their treasures looted are seldom granted much sympathy because of their perceived weakness for not defending their territory. They were not strong enough to defend themselves, and consequently deserved to lose their territory to a stronger rival. That is the basic law of nature; either defend home territory or lose it to a stronger rival, and it still rules in the modern era. The law of the strong rules the thoughts of man, and there is no sympathy for the weak.

Another important condition for the job of alpha male in the past was siring the next generation, which is also part of Mother Nature's grand design. Nothing but the best genes should be passed down to the next generation, and her system weeds out weak genes from the breeding stakes. In this way, better offspring are produced in a continuous cycle of improvement. These small and incremental changes then benefit the species as a whole as generation follows generation. The initial alpha male had a harem that lower-ranking males could not access, unless by cunning or deception by mating when the alpha male's back was turned. The harem is nature's own way of ensuring that only the leader's genes are passed down to the next generation, and still operates throughout the animal kingdom. Stags battle with horns and walruses with tusks to keep their harems, and sometimes get killed in the process. The winner then sires the next generation and passes on his strong genes.

As human evolution continued its upward trajectory, the harem system was abandoned in favour of pair- formation, with one man forming a personal bond with one woman. The past never really goes away however, and harems survive down to the present day with Arabian sheiks and religious leaders availing of many wives. These leaders believe they were created by a supreme deity in the sky who made the world and its diverse systems of life, but still practise the harem system as devised by Mother Nature. The irony no doubt goes unnoticed when they take full advantage of their high position in the harem.

In Western democracies, where monogamy is a standard requirement for the alpha male role, not every leader is faithful. Some have mistresses, and they are usually concealed from the public unless a scandal breaks out and the leader is forced to admit he has them. President John F Kennedy famously had Marilyn Monroe as one of his many mistresses, for example. Women too retain a vestige of the harem lifestyle because they are physically attracted to the leader in power. They are subliminally reminded that he is the bearer of good genes, hence the magnetic, animal attraction. The leader is always more attractive to females than the follower because he is regarded as a better guarantor of good genes, the engine that drives life. That is his appeal, and that is the reason why women in the harems of Ottoman sultans and Chinese emperors competed for their master's bed. They were hearing the ancient message of the strong leader who bore the best genes.

With the evolution from rule by brawn to rule by brain, the alpha male did not have to be the strongest member of the tribe. His primate cousins still use brawn to become the alpha male, but man abandoned brute strength in favour of brain power and that advantage spurred him to the top of the evolutionary tree. This profound change from brawn power to brain power greatly increased his mental capabilities. Now instead of being the strongest, he had to be the most cunning, using his brain to win that top spot. Here was another break from Mother Nature who requires physical strength to become the alpha male and sire the next generation. The modern leader has inherited this evolved trait and does not require physical strength to get the top job, but uses cunning to become the alpha male. He builds up alliances and cultivate allies who can help him get to the top. He is all things to all people, promising them high positions in government in return for their unwavering support. In effect, the modern political leader surrounds himself with a gang of enablers who will come to his aid in times of peril, just as the chimpanzee alpha male does.

Followers tend to tolerate flaws in the leader if he is strong, and put up with behaviour they would not tolerate in others. Primates are programmed to obey the leader and follow him, since that is the legacy of their evolution and one that is compelling. That allegiance is a deeply rooted survival instinct inherited from a primitive past. Before the discovery of agriculture, early humans lived in tribal societies that gathered and hunted, and were at the mercy of the seasons. If their territory could not support them and they were at risk of starvation, the leader invaded a territory with better resources. He led the tribe, and they put their faith in him to save their lives. The modern leader too bears the responsibility of his people, which is why they generally turn a blind eye to his foibles and misdemeanours. They tend to believe his lies and accept them as truth because they believe he has their best interests at heart. He is regarded as their father figure who shields them from a world of dangers.

Westerners who live in a democracy can expect their leaders to lie to them. They promise the stars and a bright future on the campaign trail to get elected. Lying has now become an essential part of the political job application, and accepted by the people even if they do not admit it to themselves. They generally shrug their shoulders and turn a deaf ear to the lies, in the knowledge that they part of a politician's job application. If the leader provides them with security and prosperity, the lies are usually forgotten. If on the other hand he fails to protect them or bring prosperity, the lies are invariably remembered and flung back in his face.

Lying is a legacy of the transition from brawn to brain when the potential leader had to use cunning instead of strength to get the top job. He had to promise rewards of a better future free from hunger and want to become the alpha male. He had to build up his image as a man of the people and for the people, just as the modern politician still does. No politician runs himself down or declares he is unfit or unworthy for the job. On the contrary, he portrays himself as that knight on a white horse and dressed in shining armour who will come to the rescue of the downtrodden people in their hour of greatest need.

The modern leader can be a democrat or a dictator, but his primary duty is always the same. He must offer a sense of protection for his people by having a clearly defined territory that is stoutly defended and seen to be defended. Once he fulfils that vital role, they tend to put up with anything else he does. In many countries he is above the law, even where the rule of law is applied to everyone else. There is an aura about the leader that is lacking in those who follow because he is regarded as the bulwark against danger. The people place their faith and trust in him to come to their rescue when help is needed. This is how they evolved from the past to the present, under the guidance and protection of the leader of the tribe.

Three hundred and fifty thousand years of evolution cannot simply be brushed away by modern living, nor the sophistication of modern society. Man has not reached this stage of his evolution by serendipity nor under the guidance of perceived deities in the sky whose existence are beyond the realm of positive proof. Mother Nature has written the primate rulebook for humankind to follow, and man has reached this stage in his journey through life by following that same rulebook. The latest alpha male must fulfil the obligations demanded from him by the people he leads, just as his predecessors had to fulfil their roles. His job description has not changed all that much in his short journey from the past to the present.

That is the legacy of the evolution of the alpha male.

COMPETITION FOR LIVING SPACE

Man is a territorial animal who obeys the second law, and as such must establish a territory. This imperative leads to competition that can escalate into conflicts and wars. It is worth noting that conflicts for living space and territory do not involve disparate species fighting against each other, but rivals of the same species. To put it another way, lions do not compete with leopards nor cheetahs for the same territory, but with rival lion prides. Lions and leopards and cheetahs do share the same territory, if uncomfortably, whereas rival prides cannot inhabit the same patch of ground. There cannot be two rival prides sharing the same patch of ground since neither will tolerate the presence of a rival and must come to blows. Two lion prides inhabiting the same territory is contrary to the second law of territory, and therefore cannot happen. Each pride must establish its own territory and clearly mark out its boundary by establishing scent warnings to deter intruders.

Likewise, man does not compete with other animals for territory but with his own kind, with his fellow man. Two tribes with two leaders cannot both inhabit the same piece of territory, but must fight for it as ordained by the natural law. The stronger tribe that wins the territory then marks it out with visual symbols to signal it is occupied, and that the owners are prepared to defend it by force. All tribes know the rules because they are part of the natural order, and are common to territorial animals everywhere. Humans obey the second natural law because they too are part of the natural world, and as such are bound by the laws of primate society. These laws cannot be disregarded and cannot be disobeyed since they are imprinted from birth. To be a member of the primate family is to sign up for life to the laws of that society.

At the early stages of his evolution, man gathered fruits and hunted animals within his own territory. It must be assumed that his territory was clearly marked out by visual signs that rival tribes recognized on sight, because they too were part of the same species who understood the rules. If an individual ventured past the boundaries of his own territory, he risked attack from a rival tribe who claimed that patch of land. He would have known that the territory was owned by a rival because they too would have marked out their territory. The signs of ownership were visual, whether by a river or some other feature that clearly declared it was already occupied.

The discovery of a grass that could be cultivated into wheat allowed him to abandon the nomadic gathering and hunting lifestyle to grow his own food. He tamed wild animals and changed from a hunter into a farmer and herder. He built settlements, but still required land to grow his crops and feed his livestock. The defence of his property was part of his legacy, but now he also had animals to protect. That path set him on a course against wolves who attacked his herds. The uneasy relationship between man and wolf that began when he changed from a nomad to a farmer continues to this day. The vilification of the wolf echoes down to the present day, and many people who have never seen one still fear them. The freedom from having to hunt for food to having a ready supply on his doorstep gave him time to think.

Settled societies are more advanced than nomadic ones because they are freed from having to search out food to stay alive. A nomadic society has only objective and that is to find enough food to survive, and that can be a constant quest that eats up time. The nomadic quest for food left little or no time to dwell on other pursuits, nor engage in diverse mental exercises. The settled society had time on its hands because farming had solved the food supply problem, and could turn to more intellectual pursuits of the mind. It had solved the primary law of nature by ensuring its survival. As a result of the change from nomadic to settled society, the evolution of the brain was increasingly accelerated.

The construction of permanent dwellings also required a basic knowledge of mathematics, and this activity further stimulated and advanced the brain. Living in a town or a settlement required protection from rival tribes, and that shared sense of danger bound the people together and gave them a sense of community. They had a town to protect, and invented new weaponry that could be used to repel attackers who wanted to take it over. When their settlement became larger and they needed more territory, defence changed to offence and they attacked their neighbours for more land.

As the size of the settlements increased and populations grew accordingly, more territory was needed to support them. The settlements expanded into fiefdoms and kingdoms, each with its own king or warlord ruling over his own patch of ground. This pattern is universal and has happened in every country. China had rival warring dynasties in constant warfare until a strong emperor emerged to unify the country. The intense competition between ancient Greek city states resulted in almost constant warfare for domination and territory. They had a most advanced civilization, but they were obeying the second natural law of territory. The Italian city states produced the Renaissance, but they were at each other's throats in the never-ending battles for territory. All these disparate and distant societies were hearing a siren song from a primitive and forgotten past that they were not aware existed.

The intense competition between rival kingdoms for land and survival led to the development of better weapons and tactics. Where one side gained an advantage over the other, it stimulated the rival to catch up under the fourth natural law. No competitor could be allowed an advantage that threatened the survival of a rival. The competition in ancient Greece between competing rival city states devastated the countryside and killed many thousands of warriors, but it also helped the survival of its high civilization, and of democracy itself when invaded by the Persian Empire. The advanced weaponry and tactics that had evolved by states competing with each other helped defeat the mightiest empire that the world had seen up to that time.

The evolution of city states and independent kingdoms into the nation state did not stop the competition, but instead spurred it because now the nation states would compete with each other united under a single flag. Now all its efforts could be directed into external wars, rather than engaged in internal conflicts. Nation states competed with their neighbours for more territory, which in turn led to advanced weaponry and military innovations. History repeated itself when European nations stopped the advance of the Ottoman Empire with superior weapons and tactics, and stopped its westward expansion. Without that fierce competition between European nations under the fourth law, they could not have designed better weapons and tactics.

The drive for more territory, in accordance with the second law of nature, then compelled these nations to build empires of their own. They began to explore the world, armed with new weapons that few could resist.

The competition at home had stimulated the development of superior weapons under the primary law of survival. Portugal and Spain had been engaged in intermittent warfare against the invading Moors for centuries. They built better warships to prevent the flow of reinforcements to the enemy from Africa, which were then used on voyages of exploration and conquest. Voyages of exploration inevitably are followed by wars of conquest for territory. These two small countries forged vast new empires in South America where the natives could not match the superior weapons and armour. The vast territories they acquired were plucked of their wealth, as has happened since the first leader led his troop into the territory of his neighbour.

The continental wars that Britain fought helped ignite the Industrial Revolution, which in turn enabled it to become rich and construct an empire. Empires are an extension of the second law of nature because they add more square miles to the national landbank. Its rival and competitor, France, was also building up an empire of its own, and both were in competition for the alpha male position on the continent. They became engaged in an intense competition that saw them come to blows, until Britain emerged successful. Having established its alpha male position on the continent, it could build up an empire unchallenged by its principal rival and competitor.

The two world wars that marred the twentieth century both began in Europe because that continent was the most competitive on the planet, divided on grounds of commerce, language, and nationalism. The advances in technology made by one country had to be matched by a rival under the laws of nature. The fourth law can be summed up thus: *Each gain that grants an advantage to a rival must be countered by an equal or superior gain.* There have been many theories and arguments forwarded as to their cause, with historians offering different reasons why the continent tore itself apart on two occasions.

None has examined the part played by nature because all have totally ignored the primate evolution of mankind. Historians too have split from Mother Nature and believe the actions of men can be explained by diverse factors far removed from the natural world. On the contrary, man is and remains part of nature and as such is subject to her laws. Although he believes that he controls his own destiny and makes his own decisions, Mother Nature pulls his strings.

World War One saw a cobweb of alliances across the continent that ensured it could not be confined to Europe, but would become global. This was not a struggle for small nations as the propaganda of the time suggested, nor the right of them to live in peace. Small nations do not matter in the grand scheme of things, and never have because they cannot compete for the alpha male position in the world. They have never mattered in the past and they do not matter in the present. Powerful nations are important because they are in direct competition for the top spot, and this struggle is a throwback to primate evolution when contenders jockeyed for leadership of the troop. The weakest and the smallest males could not compete for the alpha male leadership because of their lack of size and strength. The same rule applies to small and weak countries. They are not in the running for the top spot because they cannot compete with powerful nations. The best a small nation can hope for is to ally itself with a powerful nation that acts as its alpha male protector and protects it from attack.

World War One, in effect, was a bid by Germany for that exclusive position in Europe which was at the time occupied by Great Britain. Their rulers were related by ties of blood, but a more deep-seated and primeval emotion was playing out beneath the blood relationship. War could not have been avoided because there was but one way of establishing the alpha male position on the continent, and that was not by discussing it over a civilized dinner of roast beef and red wine. Both countries had vigorously competed in the arms race by building bigger and more powerful battleships. The two fleets were engaged in an arms race that made conflict not only inevitable but also predictable and unavoidable.

Planes would also play a prominent place in the conflict as each side sought to gain an advantage over its rival, just as the laws of evolution dictate. A war would be fought on three fronts for the first time in history: on land, on sea, and in the air because the weapons had been invented for these theatres. The war would spur the inventions of new weapons, and the battle tank made its first appearance during this conflict. The war could not have been avoided because there was no other way of establishing which country was the alpha male except by a trial of strength.

Some examples are worth noting from nature to serve as proof that the conflict was inevitable and could not have been avoided. When two male stags go head-to-head for the alpha male position, they size each other up for strength by strutting side by side. If one considers that his opponent is too powerful to battle, he backs off. If they are both evenly matched in size, and neither is prepared to back down, a fight is inevitable since that is the only way to prove who is the stronger. They have to fight to establish which one is the alpha male. Nature demands that they sort it out by fighting because the winner gets to sire the next generation, and pass on his successful genes. In a similar manner, when a male member of a chimpanzee troop approaches the leader, he must show his subservience. The leader stands tall and puffs out his chest, and shows he is willing to fight off the challenge to his supreme authority. If the rival refuses to back down, a fight is inevitable.

Here is the underlying reason for both world wars in Europe that has been overlooked by historians and other commentators. The conflicts could not have been avoided because neither side was in control of its actions, nor its fate. The unbreakable fourth law of competition, as Mother Nature has set down, determined their fates, and they could not avoid war. The leaders of the opposing nations were merely pawns in her great game, unable to choose their own path nor avoid conflict. The competition between the nations could not have been settled around the peace table because talking could not have decided the alpha male position on the continent. The answer could only have been found on the in the field of human conflict, and it was.

If further evidence is required that warfare follows the fourth natural law, it can be seen in military displays that are happening on a yearly basis when powerful countries jockey for the top spot in the world. When Russia holds war games with its allies, the country is exhibiting its strength and willingness to fight. When America holds war games with its allies, it too is exhibiting its strength and a willingness to fight. They are sizing each other up like two strutting stags or two male chimpanzees, neither willing to back down. They posture and exaggerate their strength by displaying the newest and most powerful weapons in their arsenals, like strutting stags showing off their impressive horns. Both Russia and the USA are currently locked into this face-off, and neither side can do anything about it because they are not in control of their actions. Both sides are adhering to the fourth natural law.

Now China has emerged to challenge for that alpha male position, and showing off its military hardware. It cannot do otherwise since it wants that top spot, just as the stag that mounts a challenge to depose the leader of the herd must show off its impressive horns. The rituals are older than time. To compete, the contender must be strong enough to depose the leader. He exhibits that strength by showing off his strong weapons. The weapons on exhibition in the modern world are nuclear, and the nation without them is not eligible to compete.

These displays of strength send out a message that the country is willing to fight for its territory if required. War games are an evolution of the natural law of competition, and a challenge for the alpha male position. The weapons on display are used to impress a rival, and to send out a warning that that they will be used if provoked. Except the world is now the stage, not a plot of land or a slice of forest. And the weapons that each side displays can also destroy the world. Man has evolved from fighting over a small patch of land in a forgotten forest in a forgotten land in a forgotten era to fighting for the world itself. That piece of real estate is the ultimate prize and always has been.

The many appeasements offered to Hitler to prevent World War Two did not succeed in preventing war. No amount of bargaining could have stopped its outbreak. The war games that he promoted with new tanks and planes sent out a signal that the country had become stronger after the defeat in the previous contest.

The need for the country to fight again can be compared to the stag that had suffered a defeat in its bid to become the alpha male. The defeated stag had retired to lick its wounds and recover its health in preparation for the next bout. It had built up its strength and become more powerful in readiness for the next trial of strength that could not be avoided. It had grown larger and sharper horns to deal with its rival, and would not back down. Germany had bigger and better weapons than before, and was ready for the second contest. There could be only one way to settle the contest for the alpha male position, and that was by fighting.

The evolution of weapons from the first crude club to the latest sophisticated inter-continental missile has been driven by the fourth evolutionary law of competition. That law, as already stated, does not pit rival species against each other members of the same species compete, and therefore man must strive against his fellow man to gain an advantage. That in turn spurs his rival to catch up or be left behind. Here is the reason for the collapse of the Soviet Union who tried to keep up with the rich West in the production of arms. Its economy was much too weak to support the effort, and led to the break-up of its empire. That too is part of evolutionary competition. They who fail to compete successfully against a rival go under.

From the ashes of its implosion, Russia emerged to again challenge for the alpha male position. It cannot but compete if it desires to go head-to-head with the

USA, or indeed China. The fierce competition is also driven by the survival instinct that is common to every species. If a rival is allowed to gain an advantage, it could use that advantage to take over the territory. This subliminal warning underpins the arms race, and always has since man fashioned his first weapon. As weapons become more destructive and deadly, they will be used because one side must strive for that coveted alpha male title. Whether or not the survival instinct prevails over the desire to be number one in the world is a question that time alone can answer. Both the need for more territory, and the need for survival, are part of the same primate legacy and neither can be easily abandoned.

If history teaches anything, it teaches that when weapons are invented, they are used. As a result, the need to gain an advantage over a competitor will continue to drive the arms race for more advanced and deadlier weapons. When one side gains an advantage, the rival side is forced to catch up, which further stokes the race. The latest phase in the ongoing competition is for hypersonic missiles to evade the enemy's defence systems. A hypersonic missile is one that exceeds five times the speed of sound. Recently Russia announced it had acquired these missiles, which triggered a response from its competitors, the USA and China.

Since an advantage cannot be tolerated by a rival, the two main rivals of Russia are said to have developed these missiles. The law of competition demanded that they must because an advantage cannot be tolerated and always provokes a direct response. The invention of a hypersonic missile triggered a response by rivals. The rival who has better weapons could conceivably take over home territory, and all the consequences that terrifying state of affairs would entail. Man is not in control of the arms race, the fourth law of Mother Nature lays down the rules. The race will continue because it cannot be stopped, nor halted in its unstoppable momentum.

The arms race can be compared to a train hurtling down a steep mountain line with no driver and no brakes. Nothing can stop its moment because the laws of physics apply to its downward momentum, and they cannot be modified nor broken. In a similar manner, the laws of nature apply to the arms race because the survival instinct supersedes every other consideration. That is the first law of nature, and by far the most important one. The need to protect home territory is hardwired in the human mindset because man is highly territorial. Territory must be protected from a takeover by a rival, whether perceived or real, and the most advanced arms are the surest and best method of achieving that defence.

A large stockpile of the most powerful weapons at the disposal of the modern leader, and they provide the best insurance policy any leader can take out. They reassure his people that they are protected under his strong leadership, and they also send out the message that he is challenging for that alpha male position in the world. No amount of peace talks or platitudes can put a brake to conflicts as they are part of the natural world where species compete with each other for valuable living space. Man cannot end the competition because he is not in charge. The laws of Mother Nature are imprinted on mankind from birth, and unlike manmade laws that can be ignored or disregarded, the four natural laws are not open to question nor disobedience. These natural laws that have taken the human race to this stage of its development cannot change or be modified because they concern survival itself, and that primary law trumps all others.

The loss of territory in the past meant the loss of livelihood, resulting in death. The mass migrations from the east that destabilised the Roman Empire were driven by peoples who had lost their territories to stronger tribes who drove them out. History teaches that the nation or a people who desires to live in peace must be always prepared for war, and willing to defend itself with every means at its disposal. It must send out a strong signal that it will retaliate with overwhelming force against a trespasser, or against an attempt at a takeover of national territory. The price of peace and security is dependent on the willingness to wage war.

The nuclear deterrent is of no use unless the enemy is made aware that it will be used. That was part of the MAD policy that kept the Cold War cold between East and West, and ensured it did not turn hot. The acronym stands for Mutually Assured Destruction, or that a strike by one side would initiate a strike from the other side. The policy was reinforced by concealing nuclear devices in underground bunkers where they could remain unseen by satellites, or mounting them in submarines. The policy ensured they could not all be knocked out in a pre-emptive strike, and the attacker would face certain annihilation. The credible threat from both sides acted as a useful deterrent.

Ironically, the development of nuclear weapons of mass annihilation helped maintain peace. Yet there can be little doubt that scientists on all sides are currently seeking to gain an advantage that gives their country a leading edge, and trigger a response from the competition. And so, the eternal arms race continues because nothing man does can alter or change the fourth natural law.

Just as germs have been a constant companion of mankind down the ages, so has war. As human populations continue to rise, wars are certain to increase both in scale and killings, and in destructive intensity. Weapon are more destructive now than at any time in history, and there are more of them around for ready cash. Governments are willing to supply them for ideological reasons, or to help their balance of payments. The war in Syria, where a violent terrorist group seized large swathes of land and carved out a new caliphate, demonstrates that not only sovereign states are ready and willing to fight for land. There are always new actors on the scene ready to take advantage of a power vacuum and seize territory for themselves. That is the name of the game, and always has been since the first conflict of troop against troop.

Islamic State carved out a caliphate across Iraq and Syria in a very short time because it took advantage of a power vacuum. Its indoctrinated disciples believed they were fighting to establish a seventh century religious state in the twenty-first century, but using tanks and rifles instead of swords and knives. They desired to create a pure society lacking in human rights and enslaving those who disagreed with their medieval ideology. Their fanaticism was fuelled by a singular ideology that regarded anyone who disagreed with it as enemies who must be destroyed. The terrorist group might claim it was taking orders from a deity in the sky, but was instead obeying the laws of nature, albeit with a savagery that only the singular ideology can inflict. The fact can't be repeated often enough that all wars are for territory, and no war has ever been fought for religion alone. The ideology has been used merely as a flag of convenience to acquire more territory. Seizing territory was the goal of Islamic State because the second natural law demands that territorial animals must have their own space.

If the first casualty of war is truth, the second casualty is refugees. The mass movements of peoples displaced by conflict has always been part of warfare, and it continues into the present day. Not every country is willing to take in refugees, and that is not because they are racist or too poor. People are motivated by their primate origins, and in times of stress or trouble they tend to fall back on the survival instinct. Territory represents a tribe or a nation of people sharing a defined plot of land that is clearly marked out by flags and other visual warnings of ownership. Refugees on a small scale might be acceptable, but on a mammoth scale are seen as something akin to an invasion force.

That perception always triggers a response.

The natural reaction to an invasion is the defence of home territory, which is hardwired into the human brain and cannot be removed or displaced, no matter what a government says or how many times it reassures its people that the refugees pose no threat. Here is the reason why some countries in Europe are reacting against mass migration from Africa. Even in the USA, which is built on immigrants, there is a mounting backlash from the refugees from South America coming in through its southern border. It should not be forgotten that refugees from the war in Syria were initially welcomed into Europe and greeted by the inhabitants. They were met with welcome signs and gifts of food by the people. There were no problems until the numbers became too large to accept, and the nightly news showed columns marching across the borderless countries of the continent. The sight of tens of thousands marching across Europe triggered a natural response that demanded the territory must be defended from the invasion force.

When politicians do not listen to the people, when they turn a deaf ear to their legitimate worries, the people will turn their backs on them. The first priority of every politician in power is to protect the people, and they who fail to perform that vital task are soon kicked out of office. The people then elect politicians who do listen to their concerns, and who do act promptly to address them. That is what happened in Europe with the election of right-wing parties who promised to end the mass migration. Then the barriers went up and the flood changed to a trickle.

There is a finite amount of land in the world, and the upward trend in human population growth shows no signs of slowing down or levelling off in the foreseeable future. The competition for territory shall continue to escalate, which will ignite more wars and create greater numbers of refugees in seek of a place to live peacefully. Wars have always produced refugees who flood across borders and they can destabilise countries not involved directly in the conflict. The war in Syria is but the latest example of a mass exodus by people fleeing for their lives to escape death. They shall not be the last refugees.

Since all wars are ignited by the second natural law, they are part of mankind's makeup. There shall always be wars because they are the fastest and most efficient way of obtaining territory, and always were. People who prefer platitude to reality and praise peace seldom consider the causes of wars. They are waged for territory, whether to expand the national holding or to retake it from occupiers who have seized it for themselves. Ask these same people who praise peace and condemn war to name a single country that has achieved its independence by peaceful means and they are left with no answer, only platitudes. Nice words might win prizes and influence polite people in the correct circles, but they solve nothing. Wars are the inevitable result of competition for territory as demanded under the second natural law. That is how the world works, and it's always better to face reality than to bury one's head in the sand and pretend the world works otherwise.

Countries must fight for their independence and to get their territory back from occupiers who have stolen it. That is a fact that nobody can deny, except the usual deniers of truth and history. Occupiers never respond to peace feelers, and listen only to the sound of gunfire and bombs directed at their rule. Territory is too important to mankind to be decided by peaceful negotiations, and this tends to happen only after a bloody war or a revolution forces the occupier to the peace table to discuss terms. No empire hands back territory without a fight, but fights tooth and nail to cling onto it for as long as possible. History points this out time and again, except history is always ignored.

War is part of the human condition because man is a territorial creature. That is an inescapable fact.

THE ETERNAL ARMS RACE

Might is right in the animal kingdom. The stag with the most powerful body and the biggest horns beats his rivals to sire the next generation. The biggest and strongest lions drive out their rivals and take over the pride and sire the next generation. The largest wolf pack takes over the rich territory of a rival pack and forces it out from its prime territory. That is the law of nature, extreme but put in place to guarantee the best genes are passed down and the succeeding generation improved. It has worked well since life was first born on Earth, and then an evolved ape broke away from the old and established order.

So, how did a frail creature with no natural defences come to dominate the world of larger and stronger animals? Man does not have claws, but yet has dominion over every animal on the planet with claws. Man does not have strength, but yet he dominates the strongest and the biggest animals on the planet. The answer, of course, lies in that remarkable asset between his ears that no other animal can match or come near. His evolved brain has given him mastery of Earth and all its life forms. That single organ has proven to be more effective than claws or tusks or strength, because man's brain invented weapons to destroy any creature on the planet.

Man was not required to evolve natural defences because he learned the art of manufacturing weapons with his brain. He did not fear predators in the night because he also tamed fire, the only animal to have accomplished this astonishing feat. He could sleep safely at night inside a cave with his family, protected by a fire and weapons. He was now the top predator on the planet, and as a result had nothing to fear from other predators. Now they rightly learned to fear him, and still do where their paths collide. There was nothing out there to challenge his domination of the natural world, except rival tribes. The fourth and unbreakable law of nature decreed that the top predator on the planet must compete with the top predator because they belonged to the same species, and the result of this competition was constant battles and warfare between tribes and nations which continue up to the present.

The development of weaponry by mankind shows a steady and upward graph as the human population increased and competition for territory grew more intense, and bloody. The competition was also spurred by the second natural law, which in turn created empires. When one country acquired an empire, its rival was compelled to build an empire of its own or be at a disadvantage. This was the reason European powers carved up Africa like gluttons at a feast, because their rivals were building empires on the continent. The weapons become deadlier and the conflicts more destructive as each side sought an advantage over its rival. The two world wars of the twentieth century were the inevitable result of the arms race between nations as each sought an advantage, and they were the most destructive wars in history.

Modern warfare is quite different from the wars waged two centuries ago, with armies facing each other on a field of battle. Now civilians are deliberately targeted for extinction as well as their homes and property. Cities are levelled to the ground and infrastructure deliberately reduced to rubble as weapons become more destructive. Hospitals and schools are targeted to reduce the enemy's willingness to fight. People are not only killed by bullets and bombs, but also from starvation, from hypothermia due to the destruction of their homes, and from the lack of good medical attention. Civilians are covertly targeted to sap the enemy of his will to fight, although routinely denied and excused by the perpetrators. The combatants usually declare that they do not deliberately target civilians and blame the other side for the atrocity. When they are found to be lying and proof is produced to show that their weapons did kill civilians, they blame it on a mistake.

The inevitable consequence of the competition between nations led to a terrible new weapon called the atomic bomb being used to wipe out the cities of Hiroshima and Nagasaki. It should never be forgotten that a weapon, once invented, is always used. History tells us that time and again, except man has always ignored history as a bad teacher. If a third world war breaks out, hydrogen bombs will be used. The world must know who holds the alpha male position, and how else can that position be determined? By the USA, Russia, and China sitting around a conference table? Talks did not, nor could not, have prevented the two previous world conflicts. The world and his wife must know which country is the alpha male, and that makes conflict inevitable.

The arms race between Britain and Germany that led to the outbreak of World War One had reached such a critical mass where it could no longer be halted. It can be compared to a chain reaction in a nuclear bomb that, once started, cannot be held in check. The weapons had to be used on the battlefield in a trial of strength to determine the alpha male position. That same critical mass is building up again as three countries jostle for that coveted top spot. Weapons can be tested in a time of peace, but no nation can be certain how they will perform in the theatre of war. There is but one way of finding out, and that is by using them against the enemy.

That mankind has been engaged in an escalating arms race since the first human fashioned a club is a direct result of the competition between the same species as required under the fourth law. The ongoing arms race has resulted in today's arsenals of mass destruction, with each side capable of destroying the world many times over. Man has used his best asset to assist him in the competitive struggle, just as other animals use their best weapon. And because the brain is man's best asset, he competes vigorously with that by ingeniously using it to gain an added advantage over the competition.

The tribe that made a bigger or a better club than the opposition could use it take over the territory of a tribe with inferior weapons. It is worth repeating that a weapon is designed with the brain before it is manufactured by the hands. The cerebral arms race literally evolved into the real thing as tribes sought to gain an advantage. The invention of the sword by one tribe led to a better sword by its rival, or a better shield for protection. Shields and armour were developed and tested in battle, as naturally as two male lions test themselves in battle to find out which is the stronger. The natural law of competition spurred the evolution of more efficient weapons. Those that were not successful were discarded, and weapons that granted an extra advantage were improved.

When fortified cities were erected with stone walls for better protection, they spurred the evolution of catapults to bring them down. Then when castles were improved by better designs and thicker walls, cannons were used to bring them down. The body protection of medieval knights evolved from chain mail to plate armour to gain an advantage, which was made obsolete by the invention of the gun. Now body armour has evolved to stop bullets, and its invention spurred the invention of superior bullets to penetrate this newest form of body protection. The eternal arms race continues to this day and it cannot be halted.

Mother Nature pulls the strings, not man.

A faster fighter plane by one nation spurs its rival to catch up or surpass the threat with a faster plane. The development of supersonic rockets by one side led inevitably to the quest for hypersonic rockets by its competitor. In the same manner, the invention of atomic bombs led to the invention of the hydrogen bomb. Then the yields or destructive power of the bombs were increased as each side sought an advantage. Here is the chain reaction that always leads to war. Man might portray himself as a rational creature, but the siren call of nature is really in control of his thoughts and actions. Man competes with his fellow man as nature has ordained, and he can do nothing to halt the competition. All the platitude and talk of a universal peace and the brotherhood of man cannot halt the arms race because competition between nations part of the natural order.

It should be noted that nations behave exactly like primates, which is hardly surprising because nations comprise primates called human beings. The present state of the world, with a resurgent Russia and its allies facing off against the USA and its allies, is a throwback to two tribes facing off against other across a river or a boundary line. Each alpha male reassures the members of his tribe that he shall protect them from danger if the rival on the other side mounts an attack. An attack on one of his tribe is the same as an attack on the whole tribe, and shall be dealt with accordingly. These are precisely the same primate rules that underpin chimpanzee society. The alpha male protects the members of his troop from a rival troop who invades their territory. This symbiotic relationship now underpins the defensive systems of both East and West.

The question has been asked consistently by pacifists and anti-war organizations why so much money is spent on arms when it could be used instead to feed the world. To seek the right answer to any question, it must be addressed to the right source. One does not examine the roots of a pear tree to discover why an apple tree is dying, but that is what anti-war campaigners always do. These groups have become so distanced from Mother Nature that they tend to seek answers in the wrong place.

There can be no answer to the arms race unless the source of human behaviour is examined. Competition between the same species is hardwired in the human brain because that is the natural law, and no force under the sun can remove the urge to compete. Even in religion which claims its divine authority from a deity in the sky, there is fierce competition. Both Christianity and Islam have been engaged in competition for fourteen centuries, and shall be until one or the other gains the upper hand. Here is the fourth law at work, a fierce competition between the two largest religions on the planet for that coveted and most desirable alpha male position. Religion too obeys the natural law because it is part of the natural world.

The struggle between tribes and nations and states and ideologies remains part of the same evolutionary process that granted mankind mastery of the planet. There can be no universal love between nations nor brotherhood of man because such aspirations contravene the natural law, and no amount of wishful thinking can change that hard fact. There has always been competition between tribes no matter in what part of the world they lived in, or the type of societies they formed. There was competition between North American tribes. There was competition between Italian cities. There was competition between Greek states. There was competition between barbarian tribes and between civilized societies. There is a pattern here that should not be overlooked, and yet always is because mankind has cut itself off from the natural world.

In obedience to the fourth natural law that cannot be disobeyed, there shall always be competition between nations. Man has channelled that natural competitive drive into sport, and the Olympic Games in ancient Greece were designed to keep the warring parties apart, at least for the duration of the games. Then the city states went back to doing what came naturally, and that was fighting each other for the alpha male prize. The modern Olympics were resurrected for the same purpose, but they could not prevent two world wars. It remains very doubtful if they can prevent a third world war either because, as stated, the arms race must produce a winner. The Olympic Games produce winners and losers, and so do wars.

Man is born to compete against his fellow man, and it can take on many forms. His nation can compete against other nations at football or on the field of athletics. Companies with similar products compete with each other both at home and abroad, each seeking that dominant position at the top. Nations likewise compete in war against other nations. They do not receive a gold cup or a trophy if they win, but they can gain that prized and coveted alpha male position. The problem with the next war is that there can be no winner, only losers.

RISE AND FALL OF EMPIRES

Empires rise in accordance with the second law of territory, and fall because they become too weak or dissolute to defend the territory they have conquered. Every empire has arisen to acquire more land, just as all wars that have ever been fought were for that most precious commodity. The secondary cause for the rise of empires is the belief in the superiority of one tribe over the other, or of one race of people over another race of people. This reason too is part of the natural order that pits one tribe against another under tribe the competitive law between members of the same species. All territorial species fight for territory against each other under the second natural law, and those not strong enough to resist a takeover lose it to a stronger rival.

Empires adhere to the principle that might is right, and as such no empire has ever been formed by peaceful means. All empires have been forged in blood since they were first created, and continue to be forged in blood since that is the ink of empire. The history of empires is not written in black ink, but in red blood. The age of empire has not died because the demand for more land is part of the human condition as territorial animals. The age of empire is merely hibernating until the right time and the right leader arrives for its bloody reawakening.

The duration of an empire depends on its strength, and on its willingness to fight for the territory it has conquered. An empire is always tested for strength by rivals, and also by bands of nomads in search of rich territory. Every empire has had to defend its borders against rivals who were also attempting to grab more land. The Roman Empire fought constant border wars against nomadic tribes who themselves had been thrown out of their territory by stronger tribes. The Persian Empire built a wall on its northern frontier to stop incursions by expansionist powers in search of new lands. The British Empire fought border wars against rebels in in India, and the Chinese Empire was under siege from nomadic tribes that resulted in the Great Wall to keep them out. The wall did not succeed, and the empire fell in accordance with the natural law.

The wars between empires were also fought for the same reason, to extend their territory by grabbing more land from a rival. The Russian Empire fought the declining Ottoman Empire and seized large tracts of territory as the latter broke up and offered rich pickings to rivals who hovered like vultures waiting to strip it clean. The Crimean War was fought solely for that purpose, to stop Russia from gaining more territory, and to carve out more for Britain and France who also wanted their share of the spoils on offer. In a similar manner, the unstable map of Europe was redrawn after World War One as the victors took their share of territory from the losers and acquired it for themselves.

This is history repeating itself time and again, with peace treaties rewarding the winners and punishing the losers. Germany was the loser; and as the instigator of the conflict was forced to pay reparations to the winners in materiel and gold. The greatest blow to its honour and sense of nationhood was dealt by the dismemberment of the country. Territory was ceded to Denmark, to Belgium, to Czechoslovakia, and to Poland. This loss of land was the unkindest cut of all, and led directly to World War Two. A nation can pay over tonnes of material annually and hundreds of gold bars to pay off reparations with no loss of face, but territory is another matter entirely. The loss of national territory remains the deepest cut that fails to heal and can be exploited by the populist leader.

There is a reason why some empires collapse suddenly and others last for centuries, and it can be found in primate evolution. The country that constructs an empire is the alpha male, as proven by the fact that it has beaten its rivals and competitors for the top spot. It then secures the top position by taking a leaf out of the chimpanzee rulebook. It actively recruits client states to its cause by offering them special advantages and privileges. These client states are the gang who support the empire, and in return are awarded special benefits unavailable to those who do not recognize the empire's right to rule. The client states recognize the empire's dominant position by paying tribute in gold or silver. Tribute has nothing intrinsically to do with money, but instead is an acknowledgement of the alpha male position.

In return for this recognition of dominance, the empire offers protection to the client states, in much the same way that the alpha male chimpanzee protects the weakest members of his troop from attack by stronger members. This essential symbiosis is vital to the cohesion of an empire, just as it is equally vital to the troop. When it breaks down and disregards the laws of nature, the empire is certain to collapse. Every empire falls from within before falling from without, and this happens when it loses its vital cohesion. Again, history provides some examples that prove this point.

Historians have long puzzled how small bands of Spanish adventurers and mercenaries could bring down empires where they were vastly outnumbered in manpower. Weapons or tactics alone could not explain how a tiny force could have brought down the mighty Aztec Empire, no matter the superiority of the weapons or tactics. They could not have done it alone as their journals suggested, and excavations are now proving that they had recruited native allies from within the empire. The Aztecs had not fulfilled their role as alpha male by protecting the conquered peoples within their empire, but had instead exploited them. Human sacrifices from the conquered peoples were offered to the Aztec gods in ritual sacrifice. When the Spanish came, they were regarded not as invaders but as liberators who would throw off the Aztec yoke. The peoples that the Aztecs had exploited joined the Spanish invaders and helped overthrow the empire.

When an empire obeys the primate code and takes care of its client states, it has a better chance of survival. By offering protection to these states, they are likely to come to its aid if attacked by an external enemy. This symbiosis is vital if the empire is to have longevity. An alliance of states is stronger than the individual state standing alone. When the empire does not follow the primate code and exploits the client states, they feel no compunction to defend it if attacked. On the contrary, they are more inclined to join the ranks of the attacker and bring down the empire. They figure that the new power cannot be much worse than the occupying power who has exploited them instead of rewarding them for their loyalty.

A similar scenario has played out in the world of crime and is happening in every major city across the world. The same rules of empire apply because primate behaviour does not change, nor cannot change. Territory is the best game in town, and the person who controls it can become the alpha male. The emperor might have a higher title and reputation than the crime boss, but the methods they both employ, are exactly the same. Every empire that has ever seen the light of day has been founded on theft of goods and property that belonged to someone else, and crime is not too different. The battles for territory between crime mobs do not merit recording by eminent historians or academics, but the same driving force spurs both emperors and crime bosses.

The territorial wars that are now being played out by crime gangs in every major Western city are for control of territory, in much the same way that empires fight for that prime commodity. The territories are clearly marked out, just as they are marked out under empires, and any member of a rival gang venturing into the territory can expect be beaten up or killed. Sometimes a truce is called, but it seldom lasts because each side needs more territory. That need is part of the human lifeforce, and crime lords are human too. No single crime boss can control one hundred percent of a city because the fourth natural law of competition inevitably throws up rivals to challenge him, just as empires are always challenged. The natural law of competition governs both emperors and crime bosses, and they must comply because it is part of their nature. The powerful crime boss who manages to take over the territories of rivals is no different in reality to the emperor who takes over countries and adds them to his empire. The crime boss recruits from his former rivals and adds their territory to his own. If he has killed the head of a rival gang, he simply replaces him with a trusted lieutenant and recruits him instead.

The conquered rivals now become part of his gang, with all the privileges and responsibilities that their position requires. They are allowed to run their own territory and make money from drugs and prostitution, but they must acknowledge the alpha male position of the boss by paying him tribute with a percentage of the takings. As already stated, tribute has nothing to do with money, but is instead an acknowledgement of the alpha male's right to rule. In the same manner, client states pay tribute to the emperor and by so doing acknowledge his right to rule.

The cohesion of the crime family depends on all its members to fulfil their obligations and responsibilities. That is how a successful empire works, and that is how a successful crime family works. If a powerful rival starts a war with the boss, his gang is expected to come to his aid and supply him with soldiers. That is exactly how empires also operate, with client states supplying the emperor with soldiers. Under the same terms and obligations, if an outside rival tries to muscle in on the territory of a member, the boss is expected to meet that threat and defend the member. A state under the protection of an empire expects the same assistance if attacked. That is the vital symbiosis that underpins every successful alliance, and it still underpins the defence systems of East and West.

When an empire is strong and seen to be strong, the client states tend to remain loyal. If the empire does not come to the aid of one of its client states in time of trouble, that is an intolerable sign of weakness. The crime boss who does not fulfil his obligations and come to the aid of a member of his gang cannot expect to command loyalty in the future. Weakness is a fault that is not forgiven in primate society, and the offended party is likely to abandon his former protector. The vital cohesion of the group is based on mutual trust, and when it breaks down the unity is shattered.

The party who is abandoned in his hour of need also questions the right of an emperor or a crime boss to that alpha male position. No empire can afford a client state to defect because it tends to encourage the others, just as no crime family can afford to have a member defect. Internal divisions are lethal to the empire and to the crime family, not only because it breaks the cohesion of the group but also because the offended party is likely to join the other side. A united empire is better equipped to resist an attack than a divided one, and the same, simple logic applies to the crime family.

Empires do not collapse peacefully but in war and in chaos. A recurring feature in the collapse of empires is that civil wars follow as rivals jockey for position in the vacuum of departed power. There can be no worse proposition for primates than a vacuum at the top, nor a society without a leader. The word *anarchy* sums up this type of situation perfectly. It literally translates as, *without a leader*, and describes what happens when the alpha male position is not established. An empire can be compared to a strong wallpaper that hides the cracks and blemishes that otherwise might be seen. The casual visitor does not see the cracks because they are covered; but when the wallpaper is stripped away, they become visible and start to crumble. The divisions in a society that the strong empire covers up only become apparent after it has collapsed or departed. Then the fracture lines of tribe and religion begin to crumble open, and war follows.

When a mob boss is taken out, there is also a war because he was the unifying force that kept the rivals under control. His removal creates a power vacuum, and primates hate that condition most of all. Every society needs a unifying force to maintain order and to impose laws, by force usually. Laws that are not backed up by force are less respected than those that carry a harsh penalty, for obvious reasons. When the new crime boss makes an example of individuals by having them killed brutally, that message sends out a stark warning to others. He has the power of life and death, just as the emperors of old had. Even crime lords must live by laws that are understood by all parties. The chaos is then put back in its box and normal business can continue as before.

The reason why both empires and crime families fight to hold onto their gains is that the loss territory is unbearable. That is the most devastating loss of all, far worse than the loss of lives or indeed family members. The casualties in territorial battles are a price worth paying, and even the deaths of family members fade with the passage of time. The loss of territory never fades from memory. As a territorial species mankind is conditioned to seize and hold territory, not to give it away or lose it to a rival. Empires do not willingly withdraw and sail into the sunset without a fight, but defend the territory from nationalists who have a better claim to ownership than they ever had. Wars of liberation have had to be fought in every country where an empire took hold, and always shall where they still exist.

When the Soviet Empire collapsed, its client states went their own way and formed new alliances. The collapse was a serious psychological blow for Russia, but not nearly as bad as the loss of territory. It is still the largest country in the world by area, though with a smaller population than Pakistan, but the loss of territory is regarded as a wrong that must be set right. Russia is now attempting to rebuild that lost empire and has already invaded Ukraine, using soldiers without official uniforms to conceal its ambitions. It matters not that the people of Ukraine want to determine their own lives and elect their own government, the recovery of territory that Russia sees as its own is driving the conflict. Here is another flashpoint that could escalate into a full-scale war, and driven by sole cause of all wars which is territory. The Russian Empire has not died, only undergoing change before re-emerging to adapt to modern conditions and attitudes.

Meanwhile, the imperial mini-wars over control of cities are being waged on a daily basis. The casualties are lower than in wars between empires, but the objective is the same. The ownership and control of territory fuels these small wars, in the same way they fuelled those great wars between empires with mass movements of casualties and human losses at an incomprehensible scale. Wars are not decided by the numbers of casualties lost by either side, but by which side won the territory. That has always been the bottom line in war, and that is what the world always remembers. The winner takes territory from the loser and the loser is forced to give it up, at least until the next war when he can recover the national loss.

The cycle never ends because it is driven by the second law of nature which cannot be ignored. That same law remains the cause of every war.

LAWS OF MAN VERSUS LAWS OF NATURE

The laws of Mother Nature are universal to all life forms, but the laws of man were drawn up to govern specific societies and peoples down the ages. A comparison is now required to demonstrate that the laws of nature are eternal, but the laws of man evolve as society itself advances and becomes more civilized. A general rule is that nomadic societies live by the laws of nature whilst settled societies form their own laws. When man abandoned the nomadic lifestyle and became a farmer and a herder, he built houses and settlements that evolved into cities. Man is by evolution a social animal who likes to live with his own kind, a lifestyle that reflects his primate origins. City living required laws on property because land was no longer held by the group as a whole, but by individuals who had pair-formed to raise their own families. Living in cities settled and civilized man, and this change needed a body of laws to reflect his new way of life. Each society drew up its own laws, and some of these are worth noting.

The widely accepted belief is that civilization first began in Mesopotamia, or the land between the rivers that is currently known as Iraq. Laws were required to stop the theft of cattle and property, as well as violent crimes like assault and murder. The guilty had to be punished to ensure that the city worked properly, otherwise lawlessness would have reigned and no city can operate where that happens. Also, by formulating laws to enable the smooth running of the city, the need for revenge was removed and reduced the violence of blood feuds. Now the city could impose justice and satisfy the offended family who had seen a loved one hurt or killed. The balance that every city and society had to find down the ages was matching the severity of the punishment to the crime. This critical and important balance between crime and punishment still rages, since it is impossible to find a happy medium that satisfies everyone.

The earliest known evidence of recorded laws, have been found in the birthplace of civilization, in the region between the rivers. Fragments of recorded laws have been unearthed baked on clay tablets written in Sumerian, the language of that area. Baked clay tablets survive because of their durability and resistance to the weather. The laws were decreed by King Ur-Nammu who lived about four thousand years ago, or two thousand years before the birth of Jesus and the reign of Emperor Augustus. Prevailing attitudes tend to influence laws, and the death penalty was decreed for theft as well as murder. Fines were also outlined for injury against the person, and the amount laid down for the loss of an eye for example.

When Babylon became an empire in that region, King Hammurabi instituted a body of laws and set them up across his territory. They were inscribed on diorite columns, a hard stone that could withstand extreme weather conditions. The laws were based on the *lex talionis* system, or the principle of an eye for an eye and a tooth for a tooth. Approximately three hundred laws were established across the empire, and dealt with property, sexual matters, criminal behaviour, divorce, and slavery. The system of law demanded that penalties must be extremely harsh as a deterrent. A son that struck his father had his hand chopped off. The penalty for incest between mother and son was burning alive for both parties.

Law also reflects the society that frames it, and Babylon was a patriarchy as every society in that era was. Husbands were allowed sexual liaisons outside marriage, but a wife who tried the same faced the death penalty along with her lover. Killing a maid-servant carried a fine, but killing a pregnant woman who was born free demanded the death penalty. The law against bearing false witness was also very harsh. The laws were severe, but they were standard throughout the empire and added to its cohesion.

Yet they were not all harsh, and some had elements of modernism that would later become part of Western secular law. The presumption of innocence was granted to the accused for the first time in history in any society. The minimum wage, which is still being debated in the twenty-first century in some democracies, was established in Babylon almost four millennia ago. Rates were specified for drivers of oxen, doctors, sailors, and many other professions. Sorcery was listed as a crime, and in turn would enter Biblical law before finding its way to Europe. The method for discovering witches under the Hammurabi code was trial by ordeal, which probably meant immersing them in water, as happened in Europe. They were weighed down with stones and thrown over a bridge, or strapped to a chair and dunked until they confessed or drowned.

Babylon was more advanced than Europe in the codification of written laws, and it was not until the seventh century BCE that the first laws were displayed on stone columns in Greece known as stele. It was not an auspicious start for the country that would later birth democracy and philosophy. The laws were drawn up by a statesman named Draco, and that is where the word *draconian* found its origins. The laws were savage in intent and application. A man caught stealing a head of cabbage was sentenced to death. Such a system could not last, of course, and the laws evolved in line with the society as it too changed. Draco's laws were eventually repealed, and in future only the most serious crimes like murder suffered the death penalty. Where there were mitigating circumstances, the perpetrator could be spared death and sent into exile.

Roman law continued to evolve to reflect changing times and attitudes, but one aspect is noteworthy. The right of appeal was granted to the accused, and that fundamental right is still exercised in modern democratic courtrooms. Another evolution of the law occurred in England under King Henry II who introduced the jury system of twelve people, or a jury of the accused's peers to render a verdict on the accused. Originally, only men were allowed to sit on juries, but they too evolved in line with the advances in society to include women jurors. The jury system is now so common that it goes without comment, but it was a radical idea at the time. The common people were now on an equal footing with judges, and could decide the fate of the accused. It was a better way of determining guilt because the law of guilt or innocence was decided by the people on a jury, and not dictated by their so-called betters.

At the beginning of the thirteenth century, the Magna Carta or Great Charter was drawn up. It has now come to represent a check on the power of kings and a declaration of human rights. These are modern ideas and bear no relation to the reason why the charter was written. The barons felt that the king was getting too powerful and feared for the loss of their lands and estates. They forced the king to recognize their rights and entitlements. The charter would eventually evolve and come to represent all classes in society, barons and commoners alike.

As already stated, law represents the prevailing attitude of its framers. The American Constitution proves this point. It was not designed to apply to all men equally regardless of colour, and its guarantee of liberty did not extend to the black slaves. Some of its framers were slave-owners, and granting freedom to their slaves meant a financial loss because they were nor regarded as fellow human beings but as property. They were not granted the same rights and equality under the constitution as white men. It required a civil war to free the slaves and treat them as human beings rather than property. Nevertheless, the guarantee of liberty did evolve to include all men and women regardless of skin colour, but it would take time.

America and Australia were used as penal colonies to send people for stealing food to feed their starving children. That no longer happens because the law evolved. Everything that lives must evolve, and that principle equally applies to the law. The evolution of law is proof of the continuous evolution of the brain, for it is that organ which conceives and shapes the law. The brains of the framer of the Hammurabi code were not was evolved as the framers of modern laws. The law has always reflected the framers who lived in that particular era, and how they thought.

There shall always be a balance in the safety of the public in comparison to the rights of the criminal. The law must be a tightrope walker who gets from start to finish without falling off. If his balancing pole is pushed too far to the right, he falls from the rope. If his balancing pole is pushed too far to the left, he falls from the rope. That is always how the law has worked, in search of that happy medium where the tightrope walker neither falls to the right nor falls to the left.

The death penalty serves as an example. It was once a common occurrence for many offences, but now has been abolished in the majority of countries. The death penalty has evolved into imprisonment for life, but the term is usually fixed and the prisoner does not die behind bars, Treason was once punished by hanging the guilty party, quartering him, and finally disembowelling him. This type of punishment is now rightly condemned as barbarism and belonging to a more savage age. The crimes that merited these punishments in the past do not merit them in the present.

It can be seen that manmade laws change over time, and those laws that do not change, such as religious laws, are backward and have no place in the modern world. They can have no future either. In contrast, the four natural laws are eternal and do not change over time. These laws that were laid down by Mother Nature have remained as constant as the North Star since she first birthed life on planet Earth in the primeval waters. There is no need for the natural laws to change, as the laws of man have done down the centuries, needing constant revision and updating to keep in line with changing thoughts and the more liberal attitudes of the people.

An example of how manmade laws change can be observed in the punishment for minor offences in the past. A few centuries ago, a man could be sent to a penal colony in Australia for stealing a loaf of bread to feed his hungry family, but a minor offence like that today would elicit more sympathy from a judge than a conviction. That severe punishment can now be seen as too harsh for a minor offence and deserving only of a caution. Modern judges would probably castigate the government for forcing a man to steal bread to feed his family. Changing attitudes demanded the law must keep in step with the changes in society. It must reflect the current mood of the people that it serves, or lose respect and is disobeyed.

The laws of Mother Nature, unlike the laws of man, are not discriminatory and apply equally to all species on planet Earth and that includes human beings. They do not advocate slavery, a practice condoned by some religions. Slavery is antithetical to the natural laws since Mother Nature provides equal opportunities for all members of the human race. Her laws do not favour one race over another race, nor one tribe over another tribe. They present a level playing field where individuals and races and tribes compete with each other. They do not promote white skin over black skin, nor brown skin over yellow skin. Unlike the recurring and divisive ideologies of mankind, they do not advocate a master race. In fact, the fourth law ensures that the emergence of a master race shall be opposed since it could grant an advantage to that race.

The four natural laws apply equally to all races regardless of colour and wherever they live. Mother Nature is truly the lawgiver who is blind to prejudice and discrimination, and who extends an equal opportunity to all peoples. No good mother treats one of her children better than another, but regards them as equal whether male or female. Each is deserving of her love and attention, each deserving of the same opportunity in life. Mother Nature is that matriarch.

The four natural laws ensure that life shall continue to evolve and improve, unless the top predator ends it with nuclear weapons. Man is quite capable of doing just that.

MOTHER NATURE SHOWS HER HAND

Territorial animals must defend their territory because its loss can result in death. If they can repel the attack, they keep control of their territory and their means of survival. If they are too weak to beat off the attack, they lose it to a stronger rival who usurps it as his own. The new owner then erases the signs of ownership and replaces them with his markings of ownership. The usual method of declaring ownership of a territory is by erasing the scent markings of the previous owner by spraying his own scent to send out a message that the territory is now his by right of victory. Any potential rival who sniffs the scent is informed that the territory is now under new control, and that the occupier is ready and determined to fight off an attempted takeover. Defend your territory or lose your territory: that simple, stark law applies to every territorial species on the planet. It is equally relevant in the present day as it was in the past, since the immutable laws of nature do not change.

A takeover by lions can be used as one example, but the same law applies to all terrestrial territorial animals no matter where they inhabit across the world.

When the dominant male lion is driven off or killed, most usually by a coalition of males, his scent is erased. The winning lions spray their own scent as a sign that the territory is now under new ownership. These scent markings inform rival prides that the territory is taken; and the scent probably contains more messages such as the strength of the new owners and their willingness to fight. In time, they too shall face a similar challenge from younger and hungrier lions as they become too old or feeble to defend their patch. The stronger genes are then passed down to the next generation. The cycle never ceases because it is part of the natural law that has ensured the survival and reproduction of lion prides.

The more man distances himself from Mother Nature, the more she shows her hand and demonstrates she is in control of his thoughts and actions. Man is a highly territorial animal, in fact the most territorial animal to have ever evolved, and he uses visual signs and signals to advertise that a seized territory is under new ownership. During World War Two when German armies overran Western Europe, the national flags of France and Holland and Belgium were replaced by the swastika, the visual sign of the new owner. National flags are a sign of ownership that states the territory is occupied by a certain nation that claims ownership. By replacing the national flags with their own, the occupiers were sending out a strong visual message that these territories were now under new ownership. The peoples of France and Holland and Belgium were too weak to defend their national territory, and it fell to a stronger power in accordance to the natural law of takeovers.

Here was scent-marking by earlier primates that had evolved into visual-marking by later and more advanced primates called human beings. Every human takeover in history has followed this basic law, and does still because it is part of the natural order. Signs of the old order are torn down and destroyed, and signs of the new order are erected to let the people know who is in charge. The Nazi swastikas and symbols of the new owner were visual reminders that the territory was no longer theirs because they were too weak to defend its integrity. The German national anthem reinforced that message of new ownership with the auditory symbol of the takeover. To ram home the message, pictures of Hitler were prominently displayed to let the people know that that there was a new alpha male at the head of their country. The peoples of France and Holland and Belgium were not ruled by a president or a queen, but by a military dictator, a strong man and epitome of the alpha male.

People must know who their leader is because that human craving is part of their evolution. The earliest humans who lived in a tribe knew who their leader was because they saw him every day and knew him by sight. The prominent pictures of Hitler informed the conquered peoples who was in control. That is the most essential facet of human rule, knowing who is at the top, and it's as vital in the present day as it was over three hundred millennia ago. As long as the swastika flew over these countries and his face was seen on offices of state, they were ruled by Hitler. When liberation finally came, the swastikas were torn down and burned along with the pictures. Then their national flags were again hoisted as a sign that the territory had been recovered.

This scene has echoed down the pages of history across the world for many centuries, and it is still happening in the twenty-first century. The names of the conquerors might change and the flags might be different, but the scene is otherwise the same. If territory is unable to be defended, it will be taken over and visual signs of the previous owner destroyed. Then the visual signs of the new owner are erected, and these need no explanation or interpretation because they are understood by everybody with a pair of eyes. The flags and symbols explain more to the occupied people than any other form of information since that is how ownership is acknowledged.

The transition from scent to sight that occurred in man's simian ancestors ensures everyone is informed about who owns the territory. Sight is a far better sense than smell for identification. A flag erected on a hill can be seen for miles, whereas scent depends on wind direction, and also locating where the scent of a rival has been deposited. And a picture is worth a thousand words that requires no translator because it explains everything the observer needs to know. The sign language of rule is a universal one understood by humans and needs no other explanation. The people know that their territory has been lost to a rival by a glance at the signs or flags flying overhead.

Religion claims it is not of this world and comes from a higher authority whose mighty will is supreme and whose knowledge is all-encompassing. Its purveyors claim that the religion has been handed down from a benign deity in the sky who is both omniscient and omnipotent, and who governs life on Earth. Their particular brand of religion was bequeathed to mankind from on high to lift people up to a higher level of happiness, and transport them to a paradise where there is no pain only everlasting pleasure. This sorrowful world is but a painful rite of passage to a better world up there in the sky. That is the final destination of sinful mankind, if only it joins up and follows their god. The story sounds convincing until examined closely, and then an old and most familiar pattern emerges.

That pattern has very recognizable traits that can easily be identified. Religion, it would appear, obeys the immutable territorial laws as laid down by Mother Nature. These laws were not handed down from on high as the purveyors of religion preach, but form part of the natural order that applies to all territorial animals. Religion too obeys those same timeless and immutable laws that all territorial creatures on the planet must follow and obey since it cannot be otherwise. It behaves no differently than takeovers by any other animal that marks out new territory. Once more, examples are used to highlight the similarities, and to prove that the rules of life on Earth do not come from deities in the sky but from Mother Nature.

The Parthenon temple in Athens was built two and a half thousand years ago to celebrate the Greek victory over the Persian Empire. It was constructed to honour Athena the patron goddess of the city, which is also named after her. It remains the most perfect temple ever constructed, and perhaps could qualify as the greatest and most stunning building in the history of architecture. Its elegant, fluted Doric columns taper as they rise to the roof so imperceptibly that they appear to be the same from bottom to top. The four corner columns are slightly thicker to counteract the atmospheric effect of seeing them against the sky. This pagan temple represents the pinnacle of classical Greek civilization. Its ruins are still impressive, and have mainly withstood the ravages of war and age.

When Christianity became the dominant religion in Greece, all traces of paganism had to be removed in line with the long-established laws of takeovers that are part of nature. The gigantic statue of Athena was removed and replaced with a cross, the visual sign of the new power in the land. Christianity had vanquished its pagan rival and acquired its territory, in in accordance to the established law of takeovers. The pagan temple was now under new Christian ownership, and the cross mounted on its roof was a declaration of that victory.

The cross was a visual symbol for the people that the temple was under new ownership. The Athenian citizens who looked up at it were informed by their eyes that the pagan temple had fallen to a stronger Christian rival.

In the fifteenth century when Islam took control of Greece, the temple was turned into a mosque. The cross was removed to be replaced by a crescent moon, the visual sign of the new owners, also in accordance to the established law of takeovers. When the people looked up at it, their eyes informed them that the Christian temple had fallen to a stronger Islamic rival. The original pagan temple, which had been converted into a Christian church, was now converted into a Muslim mosque. A minaret was built nearby to send out another visual signal that the temple was under new ownership. When Greece regained its independence, all signs and traces of the previous owner were removed and replaced with national signs.

National flags are a relatively new invention to mark out territory, and of course only came into existence when a country was unified. A single flag, in theory at least, was designed to unite disparate tribes and religions under a unified and national identity. This aspiration never fully worked as can be seen by the frictions in Belgium with Flemings and Walloons, and in Spain with Basques and Catalans who oppose rule from Madrid. Then there is the question of different religions having to live side by side, and religion is seldom compromising.

Nevertheless, the idea was a good one and intended to form a national identity wherein disparate peoples agreed to share common ground under a specific name. Every country today has a national flag as a symbol of its own particular identity, and the stronger countries have flags that are universally recognizable. Primates instinctively recognize power when they see it, so it is hardly surprising that flags of the most powerful countries are recognized by more people than flags of countries that carry no clout on the world stage.

Before flags were invented as a sign of identity, other visual symbols were applied to bind different races together under a recognizable logo. The logo of the Roman Empire was the eagle, which was carried into battle by standard-bearers, and symbolized the power and prestige of Rome. It was placed on buildings and in prominent locations to declare ownership. That symbol of power and authority was recognized throughout the empire, and sent out a visual message that Rome was in control. When the empire fell and was usurped by the Catholic Church, the eagle was replaced with the cross. This visual sign sent out a powerful message that the empire was now under new ownership, and its territories and subjects were under the control of the new power, the Church.

It is worth repeating that human takeovers of territory are marked out by visual signs and symbols, which had evolved from takeovers marked out by scent. The centuries of warfare between Christianity and Islam were all based on the natural laws of takeovers employed by territorial animals. Faith was the excuse for the wars between the religions, but the underlying cause of all wars is territory. That fact remains unchanged no matter what the protagonists on either side say or spin, or no matter the sign they carry into battle, whether secular or religious. The wars of religion between these two bitter Abrahamic rivals followed the same basic code of takeovers, which were not derived from Abraham. The mosques taken by Christians were stripped of their crescent moon signs and replaced with crosses to send out the message that they were under new ownership. The churches taken by Muslims were stripped of their crosses to send out a message that they too were under new ownership.

When the Muslim Ottoman Turks besieged and took the city of Constantinople in the fifteenth century, they turned the magnificent church of Hagia Sophia into a mosque. The crescent moon replaced the cross on the roof, the bells were melted down, the altars were smashed, and the mosaics and murals depicting Christian symbols and signs were either destroyed or painted over, thereby claiming ownership for the new power in the land. The Christian church had fallen to a stronger Islamic mosque in accordance to the natural laws. The church that had been converted into a mosque was later converted into a museum. It has since been converted into a mosque again as Turkey lurches from a secular state into a monotheistic dictatorship.

During the interminable wars between Christians and Moors in Spain over many centuries, mosques and churches changed hands frequently, always with the same result. The crosses were torn down and desecrated, or the crescent moons were torn down and desecrated, entirely depending on which side was winning at the time. The unholy warriors who fought these wars believed they were fighting in the cause of their particular god, but they were controlled by Mother Nature. When the Great Mosque at Cordoba in Spain fell to the Christians during the thirteenth century, the crescent moon on the roof was immediately replaced by a cross. The usual methods were employed to obliterate every trace of the previous owner, just as happens in nature. The mosque itself had once been a Christian church before the arrival of the Muslims. It is now a World Heritage site.

All takeovers happen in this manner and are a legacy from the past. They reveal a close connection to nature and follow the same laws that have operated since the first takeover of a forest by man's ancestors. The past has to be erased for the new regime to bed down and make the territory its own. Signs of the previous owners must be eradicated to be replaced with signs of the new owners. This immutable law is universal, and transcends both secular and religious takeovers. Mother Nature calls the shots, not mankind nor any other source.

History throws up many examples of new rulers obliterating the memory of old rulers to demonstrate that the territory is under new management. The laws of takeovers have remained virtually unchanged down to the present day, and cannot change because they are part of the natural order. Whether religious or secular takeovers, they must follow the same blueprint that pertain to takeovers of territory. The natural territorial laws cannot change because they are imprinted on the mind. Even wars of revolution and wars of independence adhere to these laws, because the blueprint has been set out since the first takeover of territory by a stronger tribe over a weaker tribe. Man is not in control of his actions, but must adhere to the natural laws that all have existed before he first evolved on planet Earth.

When the American colonies fought and won a war of liberation against the British, visual and auditory markings were established to send out the message that the land was under new ownership. The Union Jack flag was removed and replaced with the Stars and Stripes. A new national anthem was penned to signal that a new nation was born, and it was independent. By removing the flags and symbols of the colonial power and replacing them with their own, the colonists were declaring that the territory was under new ownership. A new nation had emerged from the cocoon of empire to take its place in the sun.

The French Revolution was an internal revolt by the people against the power of the ruling aristocrats and the Catholic Church, but the same unchangeable laws of nature applied. Symbols and flags of the old regime were torn down and replaced with the Tricolour flag to announce the country was under new ownership. A stirring national anthem was penned to send out an auditory message that the previous owner had been ousted and a new one installed. The French Revolution was bloodier than the American Revolution because the former was a civil war, and they always provoke the worst excesses and bloodier acts. Each side was fighting for the same territory that they both inhabited and believed was rightfully theirs, which added to the brutality and the savagery. Yet the same laws of nature applied when new states were born in America and in France, proving that the laws of nature govern the thoughts and actions of man. New flags were born in both conflicts to declare ownership of territory, and new national anthems.

The killing of the French royal family, and repeated later in Russia when the Bolsheviks killed the royal family, also have uncanny similarities with the natural world. The royal families were not only killed to prevent them from becoming a rallying point for dissidents and counter-revolutionaries, they were also killed to extinguish their genes. They died to ensure that the next generation would not carry royal genes, which were now regarded as weak genes. This subliminal warning from the past motivated and incited both French and Russian revolutionaries to act, however unwittingly. They were heeding a distant call from the natural world that only the strongest genes must be passed down. Once more, a direct comparison can be found in nature.

A coalition of lions that takes over a pride acts in exactly the same manner. If the dominant male is not killed by the invaders, he is driven out and away from the pride. He loses his territory and therefore his right to breed. To prevent his weak genes from being perpetuated, his cubs are killed before they can reach sexual maturity. It should be noted that the children of the Russian tsar were killed with the royal family, and the children of the French royal couple died before they could reproduce.

As these examples clearly show, there is no difference in takeovers whether religious or secular, nor between wars of liberation and wars of revolution. All takeovers follow the same rules and have done before recorded time, and shall in the future. The hungrier and stronger take over the territory of the fatter and the weaker. All traces and symbols of the original owners or occupiers must be erased so that the new regime can be bedded down. Their flags and visual signs are removed and replaced with the signs of the new owners. These are then displayed prominently to declare ownership of the new territory.

The ground rules have been laid down by Mother Nature and they are obeyed by mankind because they cannot be disobeyed by mankind. There is a pattern here, not a random series of events. Where a pattern exists, laws must also exist to bring them about. There is a specific set of laws about takeovers that always follows the same system, no matter where they occur or under what circumstances or in what part of the world. Mankind follows the laws because it too is part of nature. Humans cannot disobey these laws because they are not in control.

THE DOMINANT THOUGHT

Not all thoughts are equal since nature does not operate by the laws of equality, neither in body nor in mind. Instead, Mother Nature advocates competition under her fourth law, and consequently there are leaders and there are followers in all aspects of life. That competition also applies in the realm of thought. Everything that lives evolves, and thought lives and keeps evolving as new information is received. Thinking is to the mind what exercise is to the body, and every new piece of information the brain receives advances its evolution. That was how the human brain became the most advanced organ in the world, by continuously seeking and acquiring knowledge. And where no knowledge was readily available, man invented supernatural beings to explain phenomena that he otherwise could not explain.

The human mind is a warring battleground between conflicting thoughts until one comes to dominate and excludes other thoughts. That battle has been ongoing since the first tribal leaders fought for control of men's minds, and it has never ceased nor is likely to cease. Where the human mind leads, the human body follows, and consequently the thought that controls the mind can cause the body to act. The battle between Left and Right that now exists in the political realm is a war fought for control of the minds of men, just as the religious battles are fought in the same theatre. Whichever side wins the war to control the mind can then implant a dominant thought that acts as a shield to exclude any other thought to the contrary. It becomes the driving force of the person, and acts as that individual's personal god and his or her driving ambition in life.

Some examples can be used to prove this point, and also to demonstrate how thought works. If the dominant thought is money, every other thought is rendered obsolete and of no importance. It will drive a person to accumulate wealth, often with no regard for the law of the land. Great wealth usually equals greater corruption, and many fortunes are founded on the bedrock of theft and dark deeds, whether they be modern Russian oligarchs or old American oil barons. The rule of law does not matter to the man whose sole driving force is the accumulation of wealth, nor the moral codes that guide people who are not obsessed by money. A man will also betray his friends and his country for money, and his family. Enemies that engage in spy activities are aware of this weakness and will actively target individuals on the other side whose sole god is money. Everything is for sale to the man who worships at the feet of his economic god, and that dominant thought negates any twinge of loyalty he might feel for his country because it overrides everything else.

If the dominant thought is religion, he will follow that god throughout his life. The thought will not allow him to find happiness outside religion because it blocks every other avenue of escape. He might become a priest or enter a monastery to live a secluded life, or he might resort to violent ends by becoming a warrior of the faith. He can easily be indoctrinated into blowing up a plane or becoming a suicide bomber to kill innocent women and children without remorse. He has no conscience because the dominant thought is in control of his mind.

If the dominant thought is sex, he cannot hope to live in a monogamous marriage and be faithful to one partner. He must follow his god, and if that entails breaking up his family he shall do so. He can no longer control his own destiny nor his future. He cannot even consider the feelings of the people he has wronged because he is driven by his personal god. The battle in his mind has been won by that dominant thought, and he must follow its command.

If the dominant thought is football, he is not interested in any other subject. The state of the world or its many problems are of no relevance to him. He will follow his team all over the world. He cannot name the capital cities of the world, but can if his team has played there. The lives of the players on his team and how they spend their leisure team are of more importance than wars and famines raging across the globe. He is not interested on the world order, nor the tensions between nations that could end in war. His dominant thought decides what he should view and hear, and he cannot ignore the command of his inner god.

During the nineteenth century, an Austrian scientist called Sigmund Freud developed a theory called psychoanalysis. Its purpose was to find explanations for human behaviour, and at the time gained many followers and advocates. It gave birth to modern psychiatry, a practice that seeks to explore the hidden mysteries of the mind. The profession is regarded as quackery by its many opponents because the results are often not provable, which, of course, is the acid test for every theory. Anyone can propose something, but unless proven true it remains in the realm of conjecture rather than fact. Moreover, a psychiatrist can easily be fooled whereas it's more difficult to fool a medical practitioner. A person who feigns post traumatic disorder to claim compensation can easily fool a psychiatrist, but a person who seeks compensation for a broken leg can fool no doctor. His claim can be shown as false by an X-ray scan, but there is no similar device to expose the claimant for post-traumatic disorder.

One of Freud's theories was called the Oedipus Complex, based on a Greek myth which saw the eponymous character murder his father and marry his mother. Freud proposed that boys secretly wished to kill their father and marry their mother. His theory had the shock value that got him attention; but it was completely off the mark because it ignored the underlying meaning of the myth. Of much more fundamental importance, it ignored the lessons of nature that act to prohibit incest. Greek myths were moral tales for life and the Oedipus tale was, in fact, a warning *against* incest. That can easily be explained by the punishment he suffered as a result of his unnatural actions. His eyes were plucked out and he was rendered blind, thereby removing his most vital and important sense.

Here was a clear and unambiguous warning that Freud completely, and conveniently, missed. The message of the myth was a warning against incest, not a secret incestuous desire by boys. Incest deserved to be punished severely, and the severity underlined the message that the ancients would have fully understood.

Crucially, Freud ignored the laws of nature, and any theory that does that cannot be taken seriously. The proper study of the mind lies in its evolution, not in any theory that cannot be backed up with evidence. Any proposal or practice that deviates from man's origins in nature must be taken with a mountain of salt because it is based on theory alone, not backed up by evidence. Mother Nature abhors incest as much she abhors a vacuum, and that is proven time and again in the natural world. Young male lions on reaching sexual maturity are driven from the pride by the alpha male to prevent in-breeding. Matriarchs in elephant society also drive out young males for the same reason. In wolf packs, only the alpha male and alpha female breed, and males that wish to sire young must form packs of their own.

Incest is taboo in the natural world and action taken to avoid it happening.

To further disprove the theory, and also to prove that the mind is a battleground for thoughts, it shall be necessary to dissect Freud's theory and prove it as false. He divided the human mind into three parts, which he called the id, the ego, and the super-ego. The id controlled primitive emotions such as sex and survival. The super-ego was the moral conscience, and the ego acted as a sort of referee between the other two. This theory presupposes that the human mind is rational and cannot be influenced by outside factors. It also ignores the salient fact that humans are first and foremost primates. That is their heritage, and that is their legacy, and that influences how they think and how they act.

As explained earlier, the four natural laws are inherent in human babies because they have evolved that way from their primate past. They are born with the survival law, the territorial law, the procreation law, and the competitive law. Each law kicks in as the baby matures into adulthood. Babies are also programmed by primate legacy to instinctively obey the leader or alpha male. As a result, they grow into adulthood conditioned to follow the leader and can be easily indoctrinated and led. Their memory banks can be taken over and corrupted by that leader, leaving them with no free will to make their own decisions. A dominant thought can be implanted by a dominant ideology or leader that counteracts and overrides other thoughts or ideas.

A strong leader can persuade the individual that lies are truths and truths are lies and the indoctrinated person shall believe him, even if that means ignoring his own common sense. There is no ego to act as an impartial referee, nor can there be for that would suggest every person is a philosopher who can distinguish right from wrong. There is only that dominant thought to guide his way, and it does not permit any contradictory thought to enter his mind.

A common defence of Nazi war criminals was that they were simply obeying orders. That excuse was not accepted by the prosecutors, who no doubt believed everyone has free will since they had grown up in free societies. The prosecutors had probably studied Freud too, since that was the fashionable thing to do in that era. They had not been exposed to a populist leader who implanted a dominant thought that stripped them of free will. That same ideology had also subverted the legal system in Germany, and recruited the judiciary to the banner of the dictator. It had also recruited the intelligentsia to its cause, and writers who should have known better. They too were stripped of their critical faculties like the vast majority of the populace. The powerful leader who indoctrinates his followers is the leader who must be obeyed because the dominant thought ensures he cannot be disobeyed.

The war criminals had no will to resist because the ideology had removed their ability to think. They could not disobey the dominant thought because it overrode every other one to the contrary, and would not allow them to think rationally. Also, they had been conditioned by primate evolution to obey the leader who implanted the ideology. That was their heritage and that was their legacy because they had evolved from a troop with a strong leader at its head. Hitler was but the latest reincarnation of the alpha male who must be obeyed. There was no ego to tell the followers right from wrong nor good from evil, for that would suggest they knew the difference. They were not in control because the dominant thought in their memory banks told them how to behave.

It happened in Germany and it can happen anywhere, and no system of government is immune to the siren call of a strong leader. He can overthrow a democracy and the people will willingly follow and help him achieve his goal. He can abolish the instruments of the state, and they will not object. He can turn them against their own army and police, and they will go along with his policy. The rule of law or a constitution is no barrier because ultimately power resides in the hands of the people, and when he gets them onboard, he can become dictator for life in any society.

That is a salutary lesson history teaches again and again, but is always ignored because most rational people think it can never happen in their civilized country. They ignore that siren call of the past at their peril. They ignore their primate evolution, and either believe they were created by God or that the laws of nature do not apply to a civilized society. They should be reminded that the past is always present, just as childhood remains within the adult even into late age.

The natural world provides a useful lesson in the form of domesticated dogs. Although cross-breeding and selective changes have produced a dizzying variety of dogs, they all share a common ancestor in the wolf. They are said to be man's best friend and have been part of his life for a countless number of centuries. The pet owner who sits by the fireside with his dog sees a gentle and loyal companion, but he does not see the wolf inside the dog's head. He can see the body of the faithful companion at his feet, but he cannot see the mind of the dog. If that dog joins a pack at night, it will invariably revert to the wild and join the pack because it hears a siren call from the past. The siren call of its past is too powerful to ignore.

Sheep farmers who live near cities or towns have to take precautions against dogs, and erect signs that state they will be shot on sight if found near sheep. Owners of dogs shot by farmers for tearing their sheep to pieces often refuse to believe the evidence of their own eyes, and claim their gentle pet could not have been responsible. They see the body of their dead pet, but do not hear the call of the wild their pet has heard, nor the pack instinct from its past. The domesticated dog might have left the wolf behind, but the wolf has not left the dog.

Man has become civilized and sophisticated in his rise to the top, but it is a thin veneer. Scratch below the surface, and underneath lies his origins in the troop, just as the domesticated dog has its origins in the wild wolf pack. Man came from that world too and it remains inside his head. His origins lie in the troop. That is where he came from, and that is his comfort blanket which he wraps himself in when he feels threatened or in danger. The siren from the past usually comes to the surface in times of crisis, and can lead to lynching, pogroms, genocide, and revolutions. Man shows how violent and bloody he can be when law and order break down and there is no curb on his most violent and warlike tendencies. The only spark a mob needs for bloody violence and genocide is an excuse, and a strong leader to follow.

Law and order are based on consensus, a pact between the government and the people. Ultimately, the power lies with the people because they form the mob that can control the streets. If they are not kept happy and politicians are deaf to their anger, they can always revert to the siren call of the troop, and then civilization is cast aside and the mob takes control. The Roman Empire had the most powerful army in the ancient world, but the mob ruled the streets. During the French Revolution the capital was also ruled by the mob.

The siren call of the troop is a recurring theme when a crisis looms, the call from the past that galvanizes the mob into action. That same siren call has been heard many times in the past, and shall be again. It's a call to action that finds ready ears and willing followers. That siren call is part of man's primitive past that can be resurrected when he feels threatened and seeks comfort in mob rule.

The reason humans follow the primate code is no great mystery, and no theory can alter that truth. Human behaviour is based on evolution from a distinct branch of the primate family tree, and the mind evolved with the body. To examine the human mind is to examine where it came from, and its journey from the past to its at its present state of development. Just as a problem in an adult can often be traced back to a trauma suffered as a child, the worries and anxieties of modern humans can be found in their origins. To ignore where humans came from, as Freud did, is much the same as ignoring too much sugar as one of the principal causes for obesity. Again, an example is worth mentioning.

The past is never really left behind because it forms part of the journey to the present, and the mind subliminally carries warnings to aid survival. The night terrors suffered by children can never be explained by psychoanalysis, nor by any other theory not founded on evolution. Man's ancestors slept in trees by night, and cats hunt in the dark because nature has endowed them with far better night vision than primates. In Africa, primates are still hunted in the wild by big cats. Chimpanzees make their beds in trees because that is safer than sleeping on the ground. Here is the reason why children fear the dark, and that legacy also explains the night terrors they experience.

These are a subconscious reminder of a dangerous past, and triggered by the first law of survival. Man has been programmed by evolution to follow the leader, and that is how the human mind arrived to its present condition. That successful legacy is hard to shake off because it has proved successful, and also because the human mind is based on the primate model. That underpins how the world still works and how people still think. In some societies such as dictatorships, they are taught how to think because the ruling ideology does the thinking for them. They obey the ruling ideology or they end up behind bars or under the ground. In these circumstances, it is better not to think at all.

In effect, free thinking can only succeed in a free society and even then, it can be controlled. A variety of sources, such as newspapers or television stations, can offer up contrasting versions of the truth. Since there cannot be two contrasting truths in the minds of mankind, or alternative facts, the people must choose for themselves what to believe and what to disbelieve. They will invariably choose what the dominant thought tells them to believe, rather than seek out the truth for themselves. Even if the facts are explained to them by experts, they do not believe the truth because it contradicts the dominant thought that controls their mind.

What is truth? Well, that depends on who's telling it, of course. The leader of one tribe is automatically believed by his followers rather than the leader of another tribe. And since humans can serve but one leader, he is the one they lend their ears to. They close their ears to the rival leader, especially if he is telling them the truth. An example can be found in Western democratic societies where freedom of thought and expression is expected. Freedom of thought is not the same as choosing right from wrong, or lies from truth. Since they are programmed not to question the leader, they tend to believe him whether or not he tells them lies. And if the opposition candidate points out the lies, they fiercely rally behind their insulted leader as they have been programmed by evolution to do.

In a dictatorship where people are taught to think the same thoughts as the dictator, they learn to keep their heads down. They obey the primary law of nature by staying alive. That is their dominant thought and that is what they do. The majority just get on with their lives and obey the rules and survive as best they can. Only a tiny minority in every society where a dictatorship has existed revolted against the system because of the dominant thought of survival. This one thought keeps them going from day to day, and they wake up each new day with it dominating their minds.

When a populist leader comes along and implants an appealing ideology in the minds of the people, that becomes the dominant thought, and they will follow him to the ends of the Earth. The ideology does not have to be true, and indeed no ideology is ever founded on truth, but it does have to appeal to their sense of destiny, and their greatness. The ideology must have an enemy to rally the people to the leader, and there are always plenty available to choose from and vilify. An enemy provides a rallying cry for the leader, and strikes an ancient chord in the minds of his followers. They are now defending their territory from a rival who desires to take it over for himself, and such a call cannot be ignored when the message is strong enough.

An ideology based on primate evolution succeeds best of all because it appeals to the troop mentality inside mankind that has not gone away. The dominant thought implanted by a strong leader in the form of ideology can control mankind, and still does where they exist. The mind infected by ideology is unable to think for itself since the dominant thought controls it, without the person realising his best organ has been lost. There is no ego or referee to weigh up truth against lies, nor good versus evil. A theory such as Freud's can be proposed, but it should never be accepted unless it has a firm basis in fact and backed up by research. Furthermore, if it contravenes the natural laws, such as the prohibition against incest, it cannot be true. That is the final acid test of every theory.

The human mind is most susceptible to ideology, and the more extreme the better. The centre can never hope to hold against the extreme ideological pull of

Right versus Left as each side struggles in opposite directions towards its polestar dogma. The centre must eventually gravitate to one or the other because the thin fabric of moderation or centrism is not strong enough to resist the tectonic pull of either. As long as there is Right, there shall always be Left because the fourth natural law of competition instituted by Mother Nature demands a response. For every action, there is an equal and opposite reaction, and that same scientific law applies to ideology. No ideology can exist alone but must also give rise to a competing and opposing ideology in direct response.

The rise of a left-wing party in politics will always provoke a corresponding rise of a right-wing party, just as the rise of the right-wing party will be countered by the rise of a left-wing party. This natural law can be observed in the events that led up to World War Two, the worst conflict in human history.

The rise of communism in Russia at the beginning of the twentieth century inevitably gave rise to fascism in Italy less than a couple of decades later, and to the rise of Nazism in Germany. Communism could not have existed without fascism, just as left cannot exist without right. The clash between them was inevitable and predictable, and could only end in war. Both ideologies were competing to be the dominant thought across the European continent. Both desired to control the minds of its followers. That battle is still raging and shows no signs of slowing down or going away. The middle ground is always vacated because ideology by its very nature must be extreme. The extremes that communism promoted were countered by similar extremes from fascism. The ideologies could not have competed against each other otherwise. If communism had not arisen to claim the minds of men, then fascism could not have arisen to counter it.

To sum up, the dominant thought controls the mind because it drives out all others to the contrary. It renders contrasting thoughts unthinkable or conceivable. Another philosophical exercise may be used to conclusively prove this point, and to demonstrate that the dominant thought controls the human mind although it generally goes unnoticed.

Hell is regarded as a place of eternal and terrible punishment. It has entered art and literature as a place where the wicked are sent to as punishment for their sins. That concept now controls what religious followers believe about the place, and others who do not subscribe to any religious belief system. It has entered the language as a curse. *Go to Hell,* or *I hope you end up in Hell* are common phrases used against enemies. Nobody who has heard of it wants to end up in Hell.

Heaven is regarded as a place of eternal happiness where the good go to receive their rewards after living a blameless life. A tropical island paradise is often described as *heavenly.* Angels who come from Heaven are always beautiful and good. It is a place of no pain or hardship, where eternal life is granted in the presence of the creator deity. Every true believer wants to go to Heaven after death because they believe it to be a paradise in the sky. They prepare for that final journey by repenting of their sins and wiping the soiled slate of their life clean. Everybody wants to spend eternity in paradise after all.

Now, reverse the situations and propose that Hell is a place of perpetual happiness where people are granted eternal life without pain or hardship. Hell is a reward for having lived a blameless life on Earth, and where the good go after they shuffle of the mortal coil. Such a wicked idea cannot be entertained because the dominant thought prevents its entry into the minds of believers. They are unable to contemplate the ludicrous idea because it is too outrageous and defies logic. Here is but one example of how the dominant thought blocks out any concept or idea that conflicts with its control.

Let us also propose that Heaven is a place of eternal punishment where the wicked are sent to pay for their sins. Eternal fires punish the wicked for eternity and there is no escape. Such a ludicrous idea is not entertained because the dominant thought prevents believers from accepting this proposal, nor indeed considering it for a nanosecond. Heaven is a state of happiness for people who have lived a good life, and that is an established fact. That fact is not open to dispute or contradiction. Here is another example of how the dominant thought rejects any concept or idea that conflicts with its total and utter control.

There is no proof that Heaven exists or that Hell exists, but religious followers believe that they do and nothing can be said or done to change their minds. Indeed, there can be no other viewpoint but the one that has been around for millennia and has become a sacred truth that must be defended. The proposal that Heaven is a place of sinners cannot be even contemplated because the dominant thought rejects such an outrageous suggestion. In a similar manner, the proposal that Hell is a place of saints cannot even be discussed because the dominant thought rejects this ludicrous suggestion. The dominant thought controls what the person is allowed to think, and what the person is not allowed to think. Free will cannot exist where the dominant thought rules.

This primate philosophical exercise proves that the human mind does not have free will. The dominant thought controls the mind and disavows thoughts to the contrary. When a proposal or concept is unthinkable, it cannot be considered. The dominant thought does not permit free thinking when it interferes with the information stored in the memory banks. And no amount of proof can remove the dominant thought once implanted in the mind by a powerful ideology.

HOW DOGMA CONTROLS

A baby born into the world has an uninformed brain that acts like a sheet of blotting paper to absorb information. Yet the baby is already programmed by nature with an innate survival instinct, since that is the primary law. All life forms obey the primary law. A mother with a sick child will speak proudly of its enormous courage in battling illness, but so will the young of all species under the first law. The innate survival instinct is the common denominator in life. Mother Nature has not only endowed life with the primary law of survival, she gives babies that irresistible and appealing look which acts as an incentive to protect the child from danger. Having obeyed the primary law by surviving, the child is now exposed to home life.

The child's primate heritage also compels him to be highly territorial under the second natural law. His simian ancestors were born into a troop with a fixed territory, and he bears this legacy. Having left the parental home, he establishes his own territory by finding a place to live, and in the modern sense it can be an apartment or a house. Then the latent third law kicks in which is procreation and passing down successful genes. Not all children follow these laws, of course, but the vast majority do and it comes naturally. There is no need for a map to chart his way in life because he bears one from the past, from his innate primate origins. The house or the apartment is his territory, and he now can fulfil the third law by rearing a family.

The child who has become an adult has learned from his parents, and the home environment can influence the adult leaves home. The adult is then expected to think for himself and make his own decisions. If he has been indoctrinated at home, he may not be able to think rationally for himself. Or if he is exposed to a powerful and evil dogma, his mind might not be able to fight the malign influence. Often marks on blotting paper can never be erased nor removed, but remain until the paper decays and turns to dust. How this mental entrapment happens can best be explained by a philosophical exercise called the Obelisk Paradigm.

Imagine a vast desert of sand that stretches as far as the eye can see in every direction with no features except a standing obelisk of weathered granite. There are no nearby columns or monuments in the sands to indicate if it had once been part of a larger complex, or no signs to explain why it was placed in such a desolate and isolated place. The obelisk has remained untouched since it was carved, unmoving except for its shadow that circled with the rise and fall of the desert sun over thousands of years. The carved hieroglyphs have withstood the test of time and the desert sands, and are still legible.

Now, picture an illiterate farmer who has been caught in a sandstorm and stumbles across the monument. He is unable to interpret the hieroglyphs, but he understands the significance of the standing column in the sands. For him, it is a monument to the power and authority of the pharaoh who had ordained its construction. That kingly power and supreme authority came from the gods whom he would join after death. His pharaoh would be rewarded lavishly in the next life by having obeyed the gods in this life, and by keeping the land in good harmony. He departs to find his home and leaves the column standing alone again.

The second visitor to the obelisk is a lost Roman soldier who has become detached from his legion and sees the standing monument. He is unable to interpret the secret hieroglyphs, but he does understand the significance of the column. For him, it is a monument to an effete and a very superstitious people who have been conquered by a rational power from the north. His gods have human form, but the Egyptian gods have the heads of crocodiles and birds. He does not understand their obsession with the cult of death, nor the power of their priestly caste. He leaves the obelisk to re-join his lost legion.

The third visitor is a devout Christian missionary who has been fasting in the desert and finds the obelisk. For him, it is a monument to a pagan society that enslaved the Israelites and took them into captivity where they were used as slave labour to build the pyramids. They were released by the power of his god to return home and reclaim the Promised Land. His single and almighty god had proved victorious over their many pagan gods. He returns to his fasting and meditation in the knowledge that his faith alone is the only true one that exists in the world, or has ever existed.

The fourth visitor is an Islamic scholar who has taken a shortcut to join a passing caravan and comes across the obelisk. For him, it is a pagan idol and an insult to his faith. He seeks a rock to smash the obelisk, but there is nothing but sand for miles around. For him, it represents a monument to pagan ignorance that has been cast out by the coming of his enlightened faith, which is the fount of all knowledge. He rides past by to find the caravan in the knowledge that his faith alone is the only true one that exists in the world, or has ever existed.

The fifth visitor is an Egyptologist who has taken time off to answer a call of nature, and who sees the obelisk standing alone. She can read the hieroglyphs and identifies the dynasty that built it. She knows that it was not built by Israelite slaves, but by free men who were given the best medical care available. They had enough food to eat and enough beer to drink. She appreciates the form and the beauty of its art, and returns to her dig and reports the find to the authorities.

The sixth visitor is not of this planet, but an interstellar voyager from a distant galaxy. He has no prior knowledge of Earth and lands his spaceship in the desert. He disembarks and walks to examine the obelisk, but is unable to interpret the strange writing. It tells him nothing about the builders, nor why it was placed in an isolated desert. He surmises that it was built by a lost civilization who has disappeared, leaving nothing behind but a stone column for others to find. He embarks and flies to the sky, and leaves it standing alone again in the timeless sands.

Yet the column has not changed in millennia, except for the wear caused by time and wind and scorching sands. The message remains the same since the day it was written in stone. What has changed down the intervening ages is the perception of the visitors. Each one views the obelisk according to the informed knowledge stored in their memory. The Egyptian farmer has been informed of the power and might of his pharaoh who is in contact with the gods, and who keeps the world in harmony by offering them gifts. The Roman soldier has been informed by his empire, which is superior to any that has gone before.

The Christian missionary has been informed by his holy book which told him that the Israelites were captive slaves in Egypt and forced to work on her monuments. The Islamic scholar has been informed by his holy book that pagans were ignorant and their art an abomination. Knowledge has informed the mind of the

Egyptologist, and she approached the column with enough information to interpret the hieroglyphs that have remained mute since they were first carved. She alone understood the reason why it was placed in the desert and who placed it there. She alone knew the truth. Her mind has not been indoctrinated by dogma, but rendered free by knowledge.

The only visitor who had no preconception of the obelisk was the interstellar visitor. It made no sense to him and there was a good reason for his ignorance. His memory banks held no information about the obelisk. His mind could not inform him about its message for that reason.

This thought exercise explains how the human mind works, and reveals how different individuals can look at the same obelisk and see contrasting things. It explains why even in the twenty-first century some people can gaze in mute admiration at the ruins of a pagan temple, whilst others consider it an abomination and wish to blow it up. Here is the reason why Islamic State blew up pagan temples when they swept across Syria, and destroyed priceless statues and artefacts.

The eyes are the receptors for the brain which stores the memory banks of the individual. When the eyes see an image, they do not interpret the message because that is not their function. The eyes act as messengers that pass images to the brain where it is processed. The memory banks are then accessed and the answer given. Depending on the information stored in the memory banks, that is what each visitor to the obelisk saw. To put it another way, they were not seeing with their eyes but with their brains. The eyes were the go-betweens that ferried the message to the brain where it was accessed. Each brain held a different message, and the reason why each one saw something different whilst looking at the same obelisk with the same hieroglyphs.

The only visitor that the monument totally baffled was the one from interstellar space. He had no preconceived ideas because there was no information stored in his memory banks. When his eyes relayed the message to his brain, no answer was forthcoming because there was nothing to retrieve since they were blank. Memory works this way and in no other. It can be compared to how a library works. When a client visits a library and requests a book, the assistant retrieves it if the book is in stock. The client then reads the book and discovers what he or she wishes to know, and that information is stored in his or her memory banks. If the book is not stocked by the library, the client can have no knowledge of the information it contains. As a result, he or she knows nothing about the subject.

Computers act in much the same way. When a request is inputted, the computer searches its memory banks and produces a reply. If there are several sources, the replies might produce conflicting answers. Lies are stored in the computer as well as truths, not unlike the human mind. The information that a person receives when accessing the computer is not necessarily right or factual either. The seeker must interpret the conflicting answers, and if achievable sort out the lies from the truth.

Just as lies and mistruths can be stored in a computer to give out false information, the human mind can be indoctrinated to store falsehoods. Human memory banks can be indoctrinated to store lies, misconceptions, and propaganda. Once the mind has been infected by a powerful dogma, it cannot accept any information that contradicts the stored knowledge. This is the reason why many millions of people accept the creation in the Garden of Eden as literal fact. They have been indoctrinated since childhood that this creation myth is fact, and no amount of hard evidence can ever hope to change their minds.

Most people do not question the dogma because it acts as a comfort blanket, and also because they have been exposed to it when young. Young minds absorb information readily without question and usually suffer no side-effects; but if a crisis should emerge the dogma is resurrected. An attack on a sacred dogma by an outsider can trigger a response, just as an attack on territory. This is particularly relevant with a religious dogma where an attack is regarded as a personal threat. The person whose sacred dogma is attacked does not nor cannot question whether it is true or not because he is not able to rationalise the information stored in his brain. He does not comprehend that his memory banks are wrong, and have been corrupted by the dogma.

This is one reason why the Garden of Eden creation myth is widely regarded as literal truth. Followers are willingly accepting, that Earth came into existence before the sun, and that the planet somehow existed alone and without its star. Proofs are readily available that Earth could not have existed without the sun, but facts make no difference to believers in this myth. Their dominant thought rejects any proofs that contradicts the information stored in their memory banks

The followers can never come to terms with the truth that their best asset has been corrupted by the dogma, and that their minds are bound in the chains of ignorance. The dogma also ensures their brain is unable to reach its full potential because it is unable to break the controlling chains that bind it to a myth. The dogma stunts its development, just as prison stunts a person's freedom of movement.

Likewise, the ideologies of communism and fascism act in the same way. The indoctrination corrupts the memory banks, and the infected individuals do not understand that they are occupied by another force. The infected persons are no longer in control of their thoughts because their minds have been given over to the ideology. They cannot think independently and therefore are quite unable to act rationally. They stoutly defend the ideology because they believe it to be true, since that is what their minds tell them. The genocides that fascism and communism have inflicted on the world were committed by people whose memory banks were infected. They did not know the difference between right and wrong because their brains were not their own.

They were instead owned by the ideology.

THE PERSONALITY CULT

Mankind is programmed by primate evolution to follow the leader, and obey that leader. That is the ultimate legacy of the human race; and although many millennia have passed since man first evolved, man cannot escape the consequences of his origins nor his family tree. The cult of the leader still dominates how he lives, and every system of rule that ever existed has this one fundamental common denominator. It is true to liberal states and to totalitarian states, since all are evolutions of the original tribal leader or alpha male who led the tribe. The leader does not rule alone, but is aided by enablers selected from the troop or the tribe to assist him. The various systems of rule that now govern the world, whether secular or religious, adhere to this basic primate model.

In a society with a religious or secular dictator, there is no need for the leader to promote his personality since he relies on the system to keep the people in line and bend them to his iron will. Dictatorships allow for no opposition, so the leader does not require a personality cult to keep them on board. He does not have to be loved by the people, but he must be feared by them to remain in power. He creates a climate of fear by periodic displays of his absolute power. By killing his enemies or disappearing them in the night, he instils that fear factor which is essential to cow the people into submission. Fear is the best weapon ever devised for securing and retaining rule, and it still works remarkably well.

It is in the interest of his enablers to keep the dictator in power because they can enjoy the perks denied to other citizens. If a famine strikes and millions die, they will not die from hunger because they support the leader, who also has plenty of food on his table. Famines do not kill leaders after all, they kill the powerless and the poor and the weak. If sanctions are applied to the dictator, they are passed on to the common people because the rich and powerful always find a way not to suffer. The symbiosis between the dictator and his enablers is therefore mutually beneficial to both parties, and they rely on each other for mutual support.

In the liberal and democratic state, the leader has a fixed term of office before standing for re-election. The leader in this system of rule needs to develop a personality cult because he cannot use terror tactics to remain at the top. He can keep his enablers loyal by terrifying them with his personality, and by angry outbursts against them in private. The enablers are programmed by nature to take his outbursts on the chin, just as members of the original tribe had to accept a physical beating that the alpha male dished out to keep them in line. Yet the checks and balances in the democratic system places ultimate power in the hands of the people, and he must win them over if he wishes to serve more than a single term.

The natural ambition of every leader is to remain in power for life, which is not given in a democracy. Whether he serves out another term depends on the cult he has built up for himself, plus the number of his followers. If his party or enablers see that he has a large following of the people, they are likely to support him because he can win them the next election and secure their jobs. If they see he has not built up a sufficient number of cult followers, they are likely to toss him aside because he can lose them their jobs at the next election. The relationship between the leader and his enablers is based on mutual survival.

The relationship between the leader and his enablers has less to do with public service and more to do with private ambition. They cling to each other not for any outdated or personal service to the state, but to remain in power. That is the basis of every political arrangement.

No leader rules alone, and the symbiosis between him and the enablers is based on mutual dependence and survival in office. They do his bidding, and in return he awards them accordingly to keep them on board. The democratic leader rewards them with posts in his government that gives them a high profile and perks, and they remain loyal in return. If a coup is mounted against the leader, they are expected to fall in line and vote with him to see off the challenge. They depend on him for their high offices, and he depends on them to remain at the top and retain power. His position and theirs is based on sticking together through thick and thin. Everything in politics has a price, and most especially loyalty. That commodity is the most easily bought of all.

The democratic leader who casts a personality cult on the people is likely also to enrapture his enablers. He is the leader with all the answers, the captain of the ship of state who will sail them out of stormy waters and lead them into the blue horizon. They will not question his policies nor ask where he is taking the country because his strong and dominant personality has taken them captive. They too are programmed to obey the populist leader with a rainbow vision of the future where there is no rain, only bright skies and eternal sunshine. By sticking with him, they can retain power at the next election.

Power is irresistible to politicians, and in many cases more important than the destiny of a country. That is why they enter politics, for the power they crave and wish to yield. The politician in power is the man of today who makes the front pages of the newspapers and the nightly news on TV. The politician out of power is yesterday's man who does not merit the limelight of publicity. Politicians are usually more concerned with the present where they live and rule, than with the future where they do not live nor rule.

The populist leader in a democracy can extend his stay at the top by adopting policies that appeal directly to the people who voted for him, and ignoring those who did not. After all, his tribe has won the election and they should enjoy the heady fruits of victory. The tax breaks and incentives can be extended to them as payment for their loyalty and their continued support. Tax increases can be used against the people who did not support him at the last election. People who get something for nothing are more likely to vote for him again and these are his priority, not those who did not vote for him or have no intention of ever voting for him. Once he keeps the majority happy, he can afford to ignore the minority. He can then concentrate on extending his temporary power and making it permanent, and undermining the democracy.

Few if any democracies are designed to cope with the threat from within because they rely on nationalism and a sense duty to one's country to convince politicians to do the right thing. This assumption ignores the fact that the sole purpose of politics is power, and that not all politicians can be trusted to play by the rules. There are always ready ears for the populist leader with the universal message that appeals directly to the people. The man who has a way with words that appeals directly to their dissatisfaction with the current state of affairs will always find a large audience and countless ears. Democracy means nothing to him nor upholding the law. His sole aim is power, unlimited and permanent, and no democracy is safe from his siren call.

The personality cult of the leader appeals to most sectors of the community because they are programmed by primate evolution to put their trust in him. They instinctively believe him, even if they mistrust all other politicians. When he sympathises with their struggling lives, they know he is sincere, unlike other politicians who are aloof and seen only at election times. He understands the plight of the man who has been thrown out of work and forced to exist on the dole, unlike the politicians who have fat pensions when they retire.

The populist leader lends his ear to the single mother who struggles to feed her family and keep the wolf from the door. He can solve her problems because he knows how tough life can be with a child to feed. If she puts her trust in him, her child need never go hungry again. That is the message she passionately wants to hear because he is the reincarnation of that original tribal leader.

That same siren song of the populist leader applies today as it has for centuries because the requirements of human nature do not change. The people need to be reassured that they are in the capable hands of a leader who will take care of them, and fulfil their dreams and aspirations. Rulers need to be on the same wavelength as the ruled, not aloof and living in marble palaces with no idea of how the other ninety-nine per cent of the people survive. The populist leader can live in a gilded palace; but if he tells the people he's living in a shack with no roof that lets in the rain, they will believe him and see a shack with no roof that lets in the rain, not a gilded palace. He has the personality to make them see what he wants them to see.

Rule has nothing to do with reality and everything to do with perception. Once the populist leader has convinced the people that he is one of them, he's got it made. The people need someone they can identify with as their leader, not the aloof and distant political classes who rule every aspect of their lives. The populist leader is the answer to their prayers, their saviour in shining armour on a white stallion who does listen, not the politicians who turn a deaf ear to their problems. The populist leader is a man of the people and for the people. Where he leads, they will surely follow.

The ambitious leader who desires permanent power in a democracy needs only to repeat the mantra that has worked so well in the past. Enemies have to be identified and exposed before the people, both internal and external. The internal enemies are identified first to secure his power base at home, and these can be found in the opposition party. It's hardly surprising that politics is tribal because man is a tribal animal, and that entails regarding the opposition as another tribe rather than fellow countrymen and women. He then vilifies them and declares them enemies of the people, a rallying cry certain to get their backing. After all, the enemy within is always more dangerous to the safety of the tribe than the enemy at the gates. Internal enemies can sell out the people to the foe that wishes to take over their territory.

His next act is to play the grievance card, and that gets the people on board every time. The enemies of the country are responsible for all its woes, and the reason they can't give their children a good education, or put food on the table. External enemies are the reason the people are living without hope and in poverty. External enemies are engaging in unfair trade practices, and undercutting prices of homegrown products. They are putting people out of work by depressing prices. He then appeals to the people and tells them democracy is not working. What the country requires is a strong leader who can make it respected again in the corridors of power. If that mantra is repeated often enough, he will find that the people buy into it without too much thought. A simple mantra, when repeated often enough, becomes the truth.

Apathy and the general mistrust of politicians are his best allies, and the lack of knowledge about democracy. The subject is not taught at schools in the same way that language or science are taught, although there is no more important subject to be found in the history of the world. The ancient philosophers argued at length on systems of government; but that pursuit has given way to disinterest and a general public malaise with politics. The unexamined system of government is one that is unprepared for a shock to the system.

Democracy is seldom examined with a view to improvement, and to get the people to fully participate. Without democracy, the rule of law collapses and free speech is abolished. Without the rule of law, the checks are removed from man's basest instincts; and in this atmosphere the most violent man invariably arises to take control. Then the inevitable acts that have repeated themselves over centuries play out again like a recurring nightmare from the past. The rights of the individual are removed for the greater good of the state. Dissidents are led away under the cover of darkness, and never heard from again by their loved ones.

The democracy that throws up a populist leader is in danger of falling into dictatorship, and with the inevitable results. The knight in shining armour is invariably the same leader who drives the country over a cliff to drown in a stormy sea of bloodshed. History provides useful and valuable lessons that are worth repeating as an example how a populist leader can subvert a democracy and change it into a dictatorship where genocide becomes the norm. History provides an example of what can happen to a democracy when politicians do not heed the voices from the street.

History is the Cassandra of the human race. Cassandra was a Trojan princess who always foretold the truth but who was never believed. In the early decades of the last century, Germany was a weak democracy with rival factions that fought pitched battles on the streets, with the forces of law and order too weak to maintain public order. Unemployment was at record levels, inflation was running wild, and there was a general mood of despair gripping the country, a despondency that sapped the vigour of the nation. The democracy had grey politicians with no sense of direction or vision, and who offered no hope to the people. Where a political vacuum exists, it is usually filled by a strong personality who has a clear vision for the future, and a populist message that appeals directly to the masses.

The name of messenger who came to rescue Germany from its failed politicians had changed, but the same old siren sang had not changed. Hitler promised to construct a new ship of state to get the people on board. That new vessel would sail its passengers into a better world of hope and plenty. He told them that the state itself was in danger of imminent collapse, with the resultant loss of territory, the worst possible outcome for any people. Only a strong leader could save the people from their countless enemies, both internal and external. Hitler used his populism and charisma to attract a gang of enablers to his cause, the gang who would help him overthrow a democracy.

He used the classic tactics of divide and rule to gain power, and then used terror tactics to subdue and kill those who opposed his rise to dictatorship. He blamed the country's external enemies for its plight, but more tellingly he blamed its internal enemies. His message fell on willing ears, and he became leader of the country.

It did not take too long to abolish the weakened democracy and establish a dictatorship. And he rewarded the gang for their help in the form of high positions and stolen art works.

This basic system of rule, one with a leader and a gang of helpers, has operated for three hundred and fifty thousand years of primate evolution. Any system that has been so successful is not open to change, nor indeed likely to change. It underpins every aspect of rule on the planet, including the democracy. If the democracy is not strong enough to oppose the populist leader, or if the checks and balances are not strong enough to curtail the populist leader, it shall fall as surely as night follows day. Politicians who are elected to serve under a populist leader can be torn between loyalty to a country or loyalty to a leader. To put the problem in starker terms, they are forced to choose between three hundred and fifty thousand years of evolution, and democracy. It must not be forgotten that democracy is young by comparison, only two and a half thousand years in existence.

Politicians who support the populist leader are also faced with the proposition of allowing the opposition to gain power if they are defeated at the ballot box. Not all are ready to let that happen, and many undoubtedly would prefer life under their own dictator than life under a democratic leader from the opposite side. Permanent power would appear to be a better option in that scenario.

The populist leader can strip the people of their will and lead them down the path to destruction, and they will gladly follow him blindly down that path. People are primates under the skin and have evolved to live in a tribe, not independently. That is the legacy of their evolution and they cannot escape it, nor its consequences. They have reached this stage of the journey by putting a blind trust in the leader because it was his duty to protect them, and they are programmed to believe that he still does. That is the lure of the alpha male leader, and the majority take it every time, and swallow it down without realising the lure is poisoned.

A populist leader who decides to jolt a democracy into a dictatorship shall find that his gang is likely to go along with him, since they too have fallen under his spell. They shall find it hard to break out and do the right thing because of their legacy. Politicians do not exist as islands but as part of a party or a group, just like the original troop. Therefore, they behave as a troop and follow the leader. The siren song of the populist leader can prove too irresistible because it calls out from that distant past, and still sounds the sweetest music to the ears of the followers.

THE WARS OF IDEOLOGIES

Man does not wage war by bombs alone, but also uses his best asset to combat the enemy. All territorial animals use the best weapons at their disposal in the struggle for a living space. A lion uses its teeth and claws to fight its rivals for territory, and a stag uses its horns to fight its rivals for mating rights. And since man's best asset is his brain, that is the latest weapon he uses to compete against his fellow man. His brain can create an ideology that's superior to any weapon ever constructed, or indeed ever likely to be constructed. An idea can be better than the most powerful army that has ever taken the field, and more destructive than any bomb or killing device as yet devised.

An idea called Christianity succeeded where every military power before it had failed and brought down the Roman Empire from within. An idea called communism brought down the Russian Empire also from within, and with it three centuries of royal rule. In both instances, a singular ideology accomplished what no military power could ever have achieved.

No modern army is equipped to fight an idea because it has to battle against a foe it cannot defeat on a field of battle. Once a dominant thought takes hold in an army, it drives out every other thought of duty to one's country, or allegiance to any form of government. As the brain continued to evolve, ideology began to play a more important role in the struggle between tribes and nations. The brain was now being used in conjunction with the creation of weapons to beat the competition. This aspect of mind warfare was employed during one of the most devastating wars fought on European soil, and demonstrates that wars of ideology are always more destructive and barbaric than conventional ones.

The Thirty Years War was fought in the seventeenth century when tensions between Catholics and Protestants boiled over into open conflict. Both sides claimed to worship the same God, but each side believed that it was right and its rival was wrong. The dominant thought prevented either side from considering the other's point of view, of course. The singular and dominant ideology that both sides adopted could not countenance any rival ideology on the same continent. Both sides were locked into their respective ideology, and compromise was therefore out of the question. There was but one way of settling their ideological difference, and it wasn't around a peace table.

The dominant thought on both sides superseded nationalism and allegiance to one's place of birth as soldiers from the same country and city fought on both sides against their own citizens. They were united by the dominant thought that controlled their minds, not by a sense of duty to their king or country or city. Civilians were caught in the firing line as both sides tried to impose their ideology on the other by savage warfare. This is yet another aspect of the wars of ideology, the genocide of women and children because they do not think the same. The war ended in a stalemate, and the continent was divided along a religious divide. The north of the continent became mainly Protestant and the south remained mainly Catholic. But the utter devastation and barbarism caused by this crippling conflict did succeed in halting religious conflict on the continent, and led eventually to the separation of church and state.

World War Two was also mainly fought on ideological grounds, with communism pitted against fascism. Genocide was also used in this conflict as whole sections of the civilian population were marked out for extinction. The mass exterminations carried out during that conflict were a direct result of the dominant thought that stripped the perpetrators of remorse, or feelings for their fellow human beings. The enemy that does not accept the ideology of the conqueror must be eliminated. That is the bottom line of every controlling ideology, the targeting of those who do not accept its tenets. Two dominant thoughts fought for the alpha male position on the continent and it was a struggle to the death where there could be only one winner. Fascism had imposed itself on Western Europe, and communism had imposed itself on Eastern Europe. War was the only way to settle this ideological dispute, and to find out which one should rule the thoughts of the people. Just as no two tribes can share the same patch of ground, no two ideologies can coexist peacefully but must come into conflict.

After the defeat of fascism, the dominant thought of communism sought hegemony on the continent. Ideology demands that all peoples must obey its tenets and belief system. It was met by the liberal democracies that had helped defeat fascism, but this was another era and another struggle for the minds of men. The competition between East and West, between communism and capitalism in the post-war world, did not turn hot because the stakes were too high. Proxy wars were fought, but without direct confrontation between the major powers that might have resulted in a nuclear winter with unprecedented losses in human lives. Instead, each side struggled to convince members of the other tribe to see its point of view.

The clash between a dictatorship and a democracy boils down to a clash between a dominant thought on one side and none on the other. A democracy does not impose one because it permits freedom of choice, and offers an alternative to the party in power. The struggle between East and West was played out in two arenas, in the production of better weapons and in battle for minds and hearts. The confrontation between two opposing systems of rule is now known as the Cold War.

Weapons and ideology were used as twin prongs of attack to persuade the other side that it had gained an advantage. And if one side could recruit an agent in the enemy camp it could gain an added advantage to help it in the struggle. The Soviet Union recruited spies from its rival who supplied it with the blueprints to build an atomic bomb. Why go to all that expense to build the bomb when it could use the ideology to steal the bomb? That would certainly be cheaper, and it also guaranteed success first time around without having the expense of trial and error. There was no need for experimentation now that it had the blueprints for the bomb. The USA had spent huge sums of money in research and development to construct the bomb, but it cost the Soviets nothing. They simply recruited Western atomic spies to feed them with enough secret information to construct their own device free of charge and at no drain on their exchequer. Communist ideology had proved more effective in the conflict than an army of countless millions and a navy of a thousand ships.

Every ideology is specifically created to win over recruits to the cause it espouses and promotes. The stronger the ideology, the more it imprints on the memory banks and drives out other thoughts to the contrary. The ideology does not have to be true because people are more susceptible to lies and embrace them more readily than truth. That is the reason why ideology is never based on truth, because lies are always more palatable and easier to digest. Truth can be unbearable, especially when it tends to contradict the beliefs of people who have bought into the lie. Lies are readily more digestible than truth, and since everybody lies about something, truth is an unrecognized stranger in the vast majority of ears. The dominant thought or ideology then takes control of the mind and transcends national identity, or loyalty to the country.

In its more extreme form, it transcends loyalty to the recruit's wife and children, or to his mother and father. The spies who stole the West's atomic secrets and gave them to the Soviets had loyalty to no cause except the ideology that pulled the strings like a puppet-master. They bought into the ideology of a one-party state where all men and women are equal, without realizing the reality that no such society has ever existed on Earth nor is likely to exist. No meritocracy is possible in any society in any country because it contravenes the natural law of competition.

To repeat, there must be competition between species as demanded under the fourth law, with members of the same species in confrontation against each other. Under this natural system, no meritocracy can come into existence, and none ever has anywhere in the world. There must be leaders under the natural law of competition, and as a direct result there must be followers. Therefore, the primate system of rule disavows the propaganda that all men are equal under communism, nor indeed under any system of rule. The primate system of rule demands an alpha male at the top who rules the troop, and this law cannot be broken nor countermanded. That is the reason no meritocracy has ever existed in the world, at any time or in any place.

The sole purpose of every ideology is to prevent individual thought and reinforce the tribal mentality and allegiance to its tenets. When the ideology is powerful enough, it negates any sense of duty and love of one's country. This is precisely what happened to the Western spies who stole the atomic blueprints and passed them on to their Soviet puppet-masters. The ideology had stripped them of their mental faculties to think clearly, as it was designed to achieve. They had bought into the ideology of communism and a society where everyone was equal, or the Utopia that in reality was as elusive as the unicorn. The person who is not indoctrinated can observe the world clearly for himself without the influence of ideology, and observe how it works. His memory banks are not corrupted and therefore he is not blinded by ideology. It doesn't take a genius to figure out how the system of rule works. After all, this same system has been around for three hundred and fifty thousand years.

Nevertheless, the ideology worked on the spies and proved its worth. The Soviet Union was given the atomic blueprints that allowed it to quickly catch up with the rich West in its competitive struggle. It did not have to waste billions of roubles in trial and error developing the bomb. Communist ideology had handed it them on a silver platter, or perhaps a plutonium platter is a better analogy under the circumstance.

Man is primarily a creature of the mind, not of the body. His brain is his most important asset, just as the speed of the cheetah is its best asset and the tusks in the warthog are its best asset. Without his most important asset, man is as helpless as a baby after birth. The mind is to man what speed is to the cheetah and tusks are to the warthog. If a dominant thought can be implanted in his mind, it can be used to control that mind and his body soon follows. It transpires logically therefore that whoever controls the mind can control the body, and an ideology based on a dominant thought does that best. It can overcome any sense of nationalism, duty to the family, or loyalty to a friend.

As explained earlier, it can counteract facts and truth and turn them into lies. An ideology can convince people that lies are truth, and any evidence to the contrary is false. Truth is not truth unless the ideology says it is true, and anything else that claims to be true is a lie. An ideology can cause a man to betray those he loves most, abandon his wife and family, or kill his fellow citizens. A strong ideology strips away free thought from the individual and transforms him into a robot trapped in human form, willing to commit any act. That is the reason why dictators use ideology to recruit followers to their cause. It helps them in their ambition to become the alpha male, and remain there at the top.

Just as the body is prone to attack and invasion by disease or infection, so too is the human mind. The healthy human body is best suited to fight off a disease rather than the body that is weak or not resistant to the disease. The mind works in much the same way. The healthy mind that has a broad view of the world and has learned from different peoples and cultures is better suited to fight an ideology. The unhealthy mind that has not been exposed to different cultures and ideas is too weak to fight an ideology. That weak mind is easily indoctrinated, and that is why dictators always use isolation techniques to prevent information reaching their citizens. A free press or broadcasts from other countries are usually curbed by the dictator to keep the people in the dark. An ignorant people who are fed a daily dose of propaganda can be ruled more easily than an informed populace who can see past the lies cand propaganda of the ideology.

Some examples are worth noting here. During the era of the Soviet Union, the free press was abolished and the Russian people were fed a daily dose of state propaganda. They were barred from visits abroad lest they find out that their Utopia was anything but a land of milk and honey, but was instead a prison camp. Where two worlds came into direct contact in Germany, fences were erected in Berlin to prevent East Germans from visiting the West where they could observe for themselves the benefits of freedom. The fence did not work and a concrete wall was erected to stop the human exodus from the communist Utopia where everyone was equal and the leader ate the same food as the factory worker. East Germans had seen for themselves that their Western brothers and sisters had better lives and superior living conditions than they had, plus freedom of movement and no secret police plotting their movements. The ideology of communism was exposed for what it was, and that was thought control.

The same playbook is being used in North Korea to the detriment of the inhabitants. The people are fed the same ideological diet, which is probably useful because they have little else to eat. They are not allowed visit their relatives in South Korea, for that visit would expose the lies and propaganda of their divine leader, and his ideology. The freedoms and lifestyles enjoyed by their brothers and sisters in the south must remain for them an unfulfilled dream. As these two examples demonstrate, ideology not only enslaves the mind but also insists on keeping the body in perpetual captivity.

The human mind is a battleground of conflicting thoughts in which the dominant thought gains control. There is a war in the brain between thoughts for the dominant place, for the alpha male position. The purpose of every ideology is to implant the dominant thought that renders conflicting thoughts irrelevant and not worth considering. That is why dictators always ruthlessly crush dissidents and others who do not accept the ideology. That is why concentration camps were constructed in Nazi Germany and gulags were constructed in the Soviet Union. They who do not accept the ruling ideology must be punished until they see the light. They must accept the dominant thought, or face the inevitable consequences.

An example of how the dominant thought drives out other thoughts can be explained by a man who is confronted with imminent death. He might have problems in his marriage, but they are not relevant now, not at this critical time. He might be losing his job, but that's of little concern. His dominant thought concentrates his mind on one matter, and that is survival. He thinks about little else because other thoughts are excluded from his mind. They are relegated and of no importance. Money is far less important than his health and he will spend everything he has to get well again. He may not have lived a religious life, but he will try religion in the hope that it can cure him. The dominant thought of staying alive occupies his mind when going to bed by night and he wakes up with the same thought in his head. It controls his whole being, so much that nothing else matters but getting well.

Ideology in its extremist form works in this way because it is stronger than any other thought, and becomes dominant. It controls the man instead of the man controlling the ideology because his mind is too weak to fight it. Just as the weak body is more susceptible to disease, so too is the weak mind more open to ideology. This condition can be observed in religious fanatics who are under the control of ideology and who are unable to think for themselves. The dominant thought instructs them what to do, and if that instructs them to kill fellow human beings because they have a different religion or none, that is what they do. It can also be seen in the ideology that brought the Nazis to power and the death camps. The mind that is given over to any ideology no longer belongs to the individual, but to the leader who implants the thought.

Therefore, the competition between nations shall continue on two fronts, both in arms and in thoughts. There can be no centre ground because, by definition, an extreme ideology must be extremist. The Left shall always oppose the Right because the centre cannot hold against the extreme gravitational force of either. The extremes have a way of winning these mind battles because liberal thinking is not strong enough to oppose the dominant thought of either wing. The side with the stronger ideology, who can recruit from the opposing side, is destined to win the war. In the final battle, it will not be the enemy with the best missiles or planes who shall succeed, but the enemy with the best ideology. That arena has been the battleground since ideology evolved, and it shows no signs of going away.

COHESION OF THE TRIBE

The number of individuals in the first human tribe can only be surmised or approximated, but it can be stated with near certainty that it was not large. Perhaps ten to twenty adults comprised the tribe, about the size of a football team. Any larger, and the tribe would become too unwieldy and break up, losing its vital cohesion. A small tribe is easier to maintain and control than a large tribe since the alpha male can keep a better eye on the individual members and forestall a coup. That was no longer the case as the size of the tribe increased and the threats to his authority multiplied exponentially as a consequence. There is an inherent tension in the tribe driven by the struggle for the alpha male position as nature dictates, with rivals constantly in intense competition for that top spot. The dynamism of that struggle ensures that the leader is always under challenge because he must prove his fitness to rule by beating off challengers. Under the immutable law of nature that requires the best genes to be passed down, only the strongest must be allowed to acquire the top spot and sire the next generation.

A challenger to the alpha male might not accept his defeat and live within the tribe, but move away with his disgruntled supporters to form a new tribe. These fractures would have happened when the tribe became too unwieldy to control and rivals fought for the top position. This same scenario can be observed in later human societies when groups set out to form their own systems under their own leader. That is how humans came to populate the world after their initial evolution in Africa. Under primate law, no two rivals can share the same territory, just as there can be but one sun in the sky. There must be a winner and as a direct result there must be a loser. Not every defeated contender is willing to serve a leader that he does not fear or respect. Some choose to leave his territory and find one of their own where they can be the alpha male.

The first tribe was homogenous, which is to say the members were similar in appearance and came from the nearby location. They had the same skin colour and spoke in the same tongue. They knew each other by sight, and by daily interaction. They communicated with each other regularly in the same tongue, and in the process reinforced the tribal bond. The tribe was closely knit, a cohesive band that hunted together and fought side by side to defend their territory, led by the leader they knew on sight. If a member did not think in the same way as the tribe, or indeed was somehow different due to the colour of his skin, he was deemed a threat to the vital cohesion. He might be a traitor from another clan who would betray the tribe and bring about its downfall. Vestiges of this early fear can be observed in modern society, most particularly in the game of football, a game that demands the group plays and thinks as one. This team game is a throwback to the original tribe that competes against other tribes on the field of play rather than the field of battle.

What is the cause of racism? Why should it matter what colour a man's skin is, whether yellow or white or black or brown? There is an explanation for every aspect of humanity that can only be explained through the prism of nature. Racism occurs where the victims are in the minority, and the majority has usually persecuted or discriminated against the minority in every society throughout history. In the case of a football team, a player with a different skin colour can be abused by his own fans, by the same people he is playing for. He is different and seen to be different. He threatens the vital cohesion of the team because he is seen to be different. This attitude is a throwback to the original tribe when it had to defend its home territory from a takeover bid by a rival tribe. Any member who did not look the same as other members of the tribe might be a traitor. And the enemy within is more dangerous than the enemy without, and always has been.

The fear of betrayal from within can be observed in the arena of democratic politics, although the fear tends to be overlooked because modern politicians believe they are rational and civilized. They tend to forget that the political party is a tribe and they have signed up to its beliefs. Those beliefs have been promoted by their leader who is the latest incarnation of the tribal leader who ruled a patch of forest in an unknown land long ago. Modern politicians have been programmed by primate legacy to obey the leader, since to disobey is an act of betrayal against the alpha male who leads them.

That is the reason why they form a circle around the leader to protect him, often against the law. The leader can be a lawless president or prime minister; but their first reaction is to defend him from his enemies, in the same manner as the enablers of the first troop leader backed him up. A threat to the troop leader was seen as a threat to its vital cohesion, and that subliminal message remains part of their legacy. They might not hear the warning voice from the past, but it remains inside their memory and it cannot be erased. That warning voice speaks of betrayal from within, an unforgivable crime against the vital cohesion of the troop.

The recurring nightmare of the enemy within can be observed when nations go to war. Germany and Britain went to war in 1914, and law-abiding Germans living in Britain were seen as facilitators of the hateful enemy and also as spies. They had integrated into British society and established businesses; but all that changed when the nations went to war. The ancient call of the troop singled them out as the enemy within, that dreaded nightmare of the internal enemy. Now they were not part of society but the enemy within, and were attacked and their businesses looted and burned. The British royal family also had to prove where their true allegiance lay because the blood of the hated enemy coursed through their veins. King George and Kaiser Wilhelm were cousins after all, related by close ties of royal blood. The British royal family demonstrated its total allegiance to the country by changing the name, which sounded far too Germanic and traitorous for a people at war. They too had to show their colours and prove their loyalty. The ruling dynasty could not be misconstrued as the enemy within, that recurring nightmare of the tribe. Henceforth, the royal family would not be known as Saxe Coburg Gotha but by Windsor, and there are few names more British than that one.

The primate nightmare of the enemy within again reared its head in the next global conflict when Japan attacked Pearl Harbour. Law-abiding Japanese people living in America were rounded up and placed in detention centres, the majority of whom were American citizens born in the country. Over one hundred and twenty thousand innocent people were incarcerated because of that primeval fear of the internal foe. The enemy inside the gates is far more dangerous than the enemy at the gates because the outsiders are seen for what they are. The enemy within remains hidden and hard to identify, making them a greater threat than the enemy at the gates. In times of extreme crisis, mankind invariably reverts to the primate code that demands total and absolute obedience to the cohesion of the tribe.

The above examples prove conclusively that protection of the tribe from internal threats supersedes all other considerations, including human rights violations. There is more at stake than human rights when the tribe is threatened, there is survival itself under the first natural law. This subliminal message is hardwired into the human mind and forces people to react when danger looms from within. No power under the sun can counteract the warning of the threat from within.

The dreaded nightmare of members of the tribe aiding its enemy came to pass during the Spanish Civil War when General Franco marched on Madrid to overthrow the democratically elected government. It was reported that he had four columns of troops marching on the city, with a fifth column inside ready to help. Hence the term *Fifth Column* has entered the language as a group willing to undermine a society or organization from within and betray it to the enemy.

The traitors behind the gates are more dangerous than the enemy outside the gates as history teaches time and again. This primal fear is the reason why anyone who looks different or who is suspected of colluding with the enemy is regarded with suspicion when danger threatens the existence of the tribe.

Religion too offers an insight to the inherent fractures in a tribe, especially when it becomes too large. This basic rule applies to all societies in all countries and in every institution because the primate system of rule derived from a small unit of members who were familiar to each other, and who were personally acquainted with their leader who lived with them. Although religions persistently claim they are not of this world but came from beings in the sky, the fact that they behave exactly the same as every other organization proves that they are earthly. The challenge for the alpha male position is always there as decreed by nature, and the religious leader is no different than his secular counterpart. This is hardly surprising because all systems of leadership are based on the primate model, since it cannot be disregarded nor disobeyed. That same model has, after all, ensured the astonishing success of mankind, and as such cannot be easily discarded.

To examine what happens when a tribe becomes too large and unwieldy, the very early origins of Christianity provides an example. This was a minor sect within the Roman Empire that competed with many other religions for followers in a society where freedom of worship was allowed. The small number of followers bound them together in in a tightly- knit community that mirrored the early human tribe. When the religion became too large and unwieldy, it began to fracture under the increased weight of numbers. As the religion recruited more souls, its cohesion unravelled and multiple breakaway factions emerged. Then it suffered a major rupture in Europe with the Reformation that tore the continent apart and set up a rival religion that came to be known as Protestantism. This religion too fractured into many and diverse breakaway factions as it became larger.

The fractures that appear when the tribe becomes too large were manifested in Roman Catholicism when rival popes set themselves up in opposition to each other, and claimed they rightfully were the one and only true pope. Since the pope has a job for life, the only way he can be challenged for the alpha male position is to set up a rival papacy. Rival popes sprang up in Avignon in France to contest the alpha male position and claim they were the true descendants of Saint Peter, the first pope. A rival church was set up in Constantinople that claimed it was the true intermediary between the faithful and God. And because there can be no compromise in religion because of its singular ideology, the church of Constantinople split with the church of Rome, each going its own separate way. East was east and West was west, and the twain never agreed on the interpretation of the Bible.

Another way to satisfy men of ambition was for them to establish their own breakaway religion from the main body and become alpha males themselves. They could not be the pope because that position was not open for as long as the incumbent lived, but they could be the alpha male of their own breakaway sect and with all the power and prestige their position conferred. There were enough souls out there in need of salvation for every ambitious preacher to recruit to his new religion. Organized religion is more about power than spirituality, and always has been since it first evolved from paganism.

The multiple breakaway sects that religions produce demonstrate how the tribe that becomes too large breaks up and loses its cohesion. Each sect produces its own leader or alpha male, each in charge of a group of followers who regard him as their spiritual leader. There are more breakaway factions and sects within Christianity than there are days in the year, each with its own leader. There are almost as many variations in the Protestant religion as there are the in the Catholic religion, and each minor sect has its own leader. Islam also has broken into factions, and there is a major division between Sunnis and Shias. Each faction regards its rival as a heretic. Judaism has more fault lines than an earthquake zone and each produces its own leader. There can be but one chief rabbi, but anyone can break from the main religion and form his own sect.

There is no fully united country on the planet because the fourth natural law denies such an existence. The nation is a larger model of the original tribe, and therefore is more prone to fracture. These fractures are not exposed in times of peace; but in war they become apparent and shown for what they are. An external threat that unites a small tribe to defend its territory produces a different reaction in the nation because the larger numbers have weakened the cohesion. There will be hawks and doves in the nation, those wishing to use force to repel the threat ranged against the state, and those wishing to extend an olive branch to the enemy. There will be individuals who side with the enemy in the hope of saving their lives, or being offered a plum job under the new order. History abounds with men who put personal ambition above love of country.

Politics also tends to illustrate how the tribe that becomes too large leads to breakaway factions. No matter what system of rule is in place, there shall always be challenges to the leader because of primate heritage. The leader must face a challenge to prove his fitness to rule and it can manifest in many ways. Even the dictator who wields absolute power cannot afford to sleep soundly at night lest he be toppled, or murdered in his bed. He is a primate, and as such there are always challengers for the alpha male position as long as he lives. The head that wears the alpha male crown can never sleep well by night because his position is not secure. It cannot be otherwise because that is his legacy.

In the society where political freedom is granted, there is no cohesion either because the tribe has become too unwieldy. Leaders rise and leaders fall, depending on how many enablers and followers they can muster. Every party that becomes too large automatically fractures under its own weight. New parties are set up by politicians who become minor leaders, in the hope that they can achieve the top job and lead the nation as the alpha male. There can be but one leader under the primate system of rule, but by becoming leader of a smaller party the politician can gain some measure of power.

To demonstrate how the tribe that becomes too large tends to fracture under its own weight, examples can be found in every large organization on the planet. Turf wars are fought by groups and individuals who struggle against each other for power. Turf wars are fought out in the public sector and in the private sector. Turf wars are fought out in the army and in the church. They are part of life in every large organization due to the pressure of too much numbers that weaken the tight bond which holds the smaller tribe together as a unit. Turf wars are also proof of the efficacy of the fourth law. Each group and each individual compete for power and influence because they cannot afford to give their competitors an advantage.

The laws of Mother Nature that govern mankind have not changed, nor cannot change though everything else turns to dust. There shall always be one leader, and there shall always be followers. Population increases did not alter this immutable law, but instead saw the larger tribes, fracture and break asunder. The leader at the top shall always have to face a challenger for that is how the system works and has done since his ancestors invented politics. Democratic politics too follows this same arrangement, with an alpha male leader at the top and enablers to maintain him in power. They are also there to stab him in the back at the most opportune moment should he demonstrate weakness or lose the people. The democratic leader does not know all his followers, but he does know his enablers because they are a cohesive band. They are the modern equivalent of the helpers who aided the first leader of the original human tribe.

As already stated, the original tribe was small and led by a leader who knew every member. This system had to be modified as populations grew and the leader could not possibly know each member of the tribe. This looser arrangement allowed smaller sects to emerge, each with its own leader. The rifts that occurred threw up more tribes who had to move away, since no two leaders can share the same piece of land. The breakaway tribes then established independent territories, each with its own leader. An example of breakaway factions can be found in the multiplicity of Native American tribes that all derived from a single ancestor tribe. Each tribe had its own leader or chief.

The tribe that reaches a critical mass of members has to fracture because it loses cohesiveness. That element is the glue that binds it together. A modification was required to deal with larger tribes and to prevent factions breaking away and losing numbers needed for defence of territory. It came about by the creation of minor leaders within the tribe who could command their own sects. In return, they had to recognize the alpha male as overall leader. The formation of the nation state is a prime example of this modification to bind rival factions together. The unification of disparate tribes and sects under a single leader allowed minor leaders to command their followers, under the condition that they paid allegiance to the leader. Feudal Europe operated under this system, with barons and warlords owning armies and territories, but recognizing the supreme authority of the king as alpha male. They were allowed to maintain private armies and keep their titles and lands, in return for acknowledging the king's divine right to rule as the alpha male.

This adaptation to larger human populations now governs all aspects of life in every country on the planet, and also every organization. When a company becomes larger and recruits more staff, new layers of managers are required to handle the increased numbers of employees. The boss no longer is personally acquainted with all the employees, and each manager is responsible for a group of employees who regard him or her as their leader. In turn, the manager pays allegiance to the boss as the alpha male of the operation.

This same hierarchal system now governs the world of business, the church, and every other institution. The factory with a small number of employees requires only one boss to run it because everyone knows the boss. Conversely, the factory with a large number of employees does require a layer of bosses because the boss does not work at their side. The basic model of governance however has not changed, but evolved to meet the challenges posed by the larger tribe. There still can be only one alpha male in every organization as the primate model demands, just as there can be but one head of a chimpanzee troop. Mankind cannot serve two masters because the primate system of rule is predicated on a single leader that it recognizes as the sole alpha male.

That fundamental law is not open to contradiction because it remains part of evolution. The invention of minor leadership roles allowed for upward mobility to encompass the larger tribe as populations grew exponentially. How the system operates can be observed in the military. A soldier begins his career as a lowly recruit who obeys the orders of a sergeant. As he moves up the ranks, he takes on responsibility for the soldiers under his command. Each layer has its own title, and that defines where he fits in the chain of command. Yet every title in the army recognizes the general with the most stars as the alpha male.

Churches that claim they are created by an omnipotent deity in the sky behave in precisely the same way. The priest starts at the bottom before ascending the ladder of command to become a bishop. He now has priests and deacons under his command. Then he rises to become a cardinal, and has bishops under his command. Each step he takes on the ladder to the top grants him more power, and more authority. His higher position and rank in the hierarchy is illustrated by his title, and by the clothes he wears. Yet he still recognizes the supreme authority as the head of his church, the alpha male who must be obeyed. This stratified system is no different to any other chain of command, either in business or the military. Its origins lie in the necessity of man's ancestors to maintain numbers within the tribe and prevent breakaways. A larger tribe offered a better defence of territory than a small tribe.

Royalty is also based on this modification to cope with the strains caused by larger tribes. There can be but one king, but there is a hierarchy of titles under him, each with its own unique title. As already stated in this work, the primate tree does not grow royalty and consequently these titles are a human invention to keep power within a certain class who believed they had the divine right to rule granted from a deity in the sky. The titles are too numerous to mention and their order of preference too arcane to understand; but they do follow the same modification as applied in every other institution on the planet whether secular or religious. None except the king is permitted to wear the crown, and each title fits in somewhere below him in order of designated title. The title is a ticket that allows the bearer is occupy his or her position in the royal hierarchy. The king is the alpha male whose high authority is recognized by every title below him.

The establishment of tribal hierarchy by bullying in the past evolved in line with the development of the human brain. It did not end the practice, which still occurs because it is part of nature, but it did allow adults to fit into society. The invention of titles and ranks created a system whereby everyone knew their place and allowed the tribe to cement its vital cohesion. As a result of the hierarchal system, the children born into a privileged class had no need to establish their own place in society. They inherited their roles by right of birth. The sons and daughters of kings and queens were born into the titles of princes and princesses, thereby establishing their high positions in the pantheon of royal hierarchy.

This unique inheritance is still happening with royal families in the modern world. The sons and daughters of kings and queens are automatically assured of their place in society, and have no need to compete with others to find their place in that society. Even in modern democratic societies, princes and princesses have a higher status than other citizens as a cursory glance at the media proves. They are invariably portrayed as superior beings who are above the common throng of humanity.

Hierarchy happens in business where the son or daughter of the owner of a company has privileges not available to fellow workers. The son or daughter might start on the bottom rung of the company, but that is simply a gesture to give the illusion that all workers are equal. Sooner rather than later, that son or daughter will ascend the ladder to become the alpha male. His or her rise has nothing to do with merit nor qualifications for the top job. The top position has already been ordained under the hierarchal system that first evolved to accommodate the larger tribe. The creative brain of man has solved the problem of exploding numbers by creating a system that appears to be meritorious, whilst acknowledging there can be but one alpha male. This system now applies to every organization in every country on the planet, whether religious or secular.

THE CONFUSION OF LANGUAGE

Language evolved under the first law of survival as a defence against predators, before developing into the principal tool of communications used by humans across the world. Vervet monkeys still communicate by verbal warnings to warn the troop of the approach of a predator, and scientists have isolated three distinct calls that bring an instant reaction. The warning call for an eagle sees the troop turning their eyes to the sky and seeking refuge in the trees where they are less vulnerable to aerial attack. The warning call for a leopard sees them scatter to the outer branches of a tree where the cat is too heavy to reach them. The warning call for a snake produces a different reaction. They do not flee immediately, but identify the location of the predator. The snake has then no chance when it loses the element of surprise and the monkeys are safe to watch it slink away.

All living things evolve, and language is no different because it too lives. An example can be found in Latin, which did not die after the fall of the Roman Empire as widely believed, but changed to meet the new conditions on the ground. It evolved into separate and distinct tongues that are now known as Romance languages. It also helped shape English which borrowed many words that have become part of everyday speech. Latin phrases are used in the courts of law, to impress juries that the lawyers are smart and educated rather than for legal reasons because there more than enough English alternatives available. The question must be posed as to why Latin evolved? Why did the language evolve to form separate and distinct dialects in Italy, Spain, and in France? This work shall now propose a theory to explain why it evolved, and offer proofs to back it up.

No theory should be drawn out of thin air unless supported by a body of evidence to give it credibility. Language changes because the fourth law of competition between the same species does not allow different tribes to speak in the same tongue. Language unites a people, but crucially it also divides them and separates them into distinctive groups. A tribe or a people speaking in the same tongue were united by that common language, which acted as a glue to bind them together in a closely-knit community. It was a common bond of their identity, a badge of belonging to the same group or tribe. They were different to the tribe or people who lived on the other side of a boundary river who spoke in a different language. They could not communicate with each other because they did not share a common tongue; and it is always easier to mistrust and of course to misunderstand those who do not speak the same language. They are strangers, not one of us, not part of our tribe.

This scenario would have been the case when tribes competed fiercely with each other for territory, with no means of common communication because they could not speak each other's language. It should be noted that when Latin ruled the Roman world, the traveller from Spain could visit Italy and be understood. He could travel from North Africa to Britain and be understood in both societies and during the journey. There was one alpha male ruling the empire and one language uniting that empire. Latin was the common glue that bound the peoples of the empire together. They were part of the same tribe, united under by sharing the same leader and the same language. After the fall of the empire, Spain and Italy evolved their own language, derived from Latin. So too did Gaul, later known as France. These countries were now competitors under different alpha males, each now in competition because they were disunited. They were now divided by different languages whereas before they had been united under a single and binding one.

From primitive warning calls to signal the approach of predators, language became the sole means of communication. As man's ancestors evolved, language of necessity became more diverse and of greater importance. There would have been no need for a complex language under the rule of brawn when the alpha male reigned by brute strength alone, as in gorilla society. It was essential to evolve complex language skills when man changed from rule by brawn to rule by brain. Now the tribe had to be governed by words, and the leader with the most persuasive argument and best words could command them. Mankind is still led by words and fine speeches; and great orators have always moved the people throughout the course of human history. Orators like Martin Luther King are admired internationally and their speeches referenced in debates. The power of good and noble words to move people are as compelling to modern ears as they were to ancient ears.

Conversely, the populist leader can sway them to his cause by appealing to their emotions through their ears. He does not need an enforcement squad to get their support, but his rhetoric. That was how primitive leaders won them to his cause, and that is how modern leaders win them over. Mankind has consistently been led by fine words and compelling rhetoric, and always shall be because of its evolution. Hitler's oratory had a spellbinding effect on the German people and stripped them of their critical faculties, and their sense of right versus wrong. Words are the gateway to the mind and can be used to implant an ideology that holds the victim captive.

Language also follows the immutable laws of evolution. When the strong language collides with a weaker one, the stronger survives and the weaker is gradually rendered extinct. Some words are usually absorbed into the stronger language from the defeated one, which makes the conqueror stronger still because it has a broader base and appeal. Incidentally, this is precisely what happened when monotheism eradicated pagan religions. It absorbed some acceptable pagan practices and made them its own, thereby broadening its appeal.

For example, the feast of Christmas was originally a mid-winter pagan festival to mark the turning point of the year. The feast of Easter too was a pagan festival to celebrate the beginning of spring and the rebirth of life after winter. By absorbing pagan practices and making them its own, Christianity became more powerful by gaining more recruits from the pagan population.

The languages that the builders of empires brought alongside their weapons encountered many diverse dialects, and rendered them extinct. Spanish eradicated the indigenous languages it encountered in South America, and Portuguese wiped out the languages it met that were spoken by the tribes in Brazil. These two European languages now dominate South America and have become the official ones governing that continent. Spanish is now the principal language in the majority of South American countries, and that language evolved from Latin which had once ruled in Europe and North Africa.

English too wiped out weaker languages that it encountered, but also absorbed many new words into its vocabulary from the weaker tongues. The contacts between other languages through trade and conquest not only enriched it but made it more global and more popular. Insular languages with no outside contacts do not advance the evolution of the human brain, but languages that travel do spur its development. For example, tribes in tropical regions that had never seen snow would have no word to describe ice crystals falling from the sky and turning the land white. They could not form a word to describe an event that they had not encountered and were unaware existed.

A global language that encountered new sights and unknown events did have to invent a new word, or borrow an existing word from the defeated language. One word out of countless thousands can be used to illustrate this point, and it is *tattoo*. This is a Polynesian word that was directly incorporated into the English language through trade and conquest. Each new word that a language creates not only enriches the vocabulary, it *crucially* also forms a new neuron connection.

Each new neuron connection that is made further advances the evolution of the brain. There is a saying that travel broadens the mind, and it does because every new sight and novel experience stimulates the upward trajectory and development of the brain.

The empires that arose under the second natural law were united under a single tongue. That common language made trade easier to conduct, and laws easier to administer. The language also united the peoples it ruled, and gave the secure sense that all belonged to the same tribe. Belonging to a larger tribe gives the added bonus of that vital sense of security that is essential to the mental welfare of humans. When empires ruled peoples tolerantly, they were generally happy with their lot and content to live under its benign rule. The empire offered them that essential need of protection, and they could rise to the top by adopting the strong language of the rulers. They could gain high positions in the ruling establishment by speaking in the same tongue. Therefore, it was a smart move to learn the language of the conqueror. That was how Britain ruled the vast subcontinent of India, by local civil servants trained in the language of conquest. It did not require a huge army to administer the huge country, but an army of local civil servants versed in English.

When empires collapse through wars and new states emerge, barriers are erected in line with the first law of protection of life. The new state protects itself by retreating behind defensible barriers and prepares to repel an invasion. National territory must be protected at all costs, and signs and symbols erected to send out a warning that the land belongs to the nation. By retreating behind barriers like mountains or rivers, the new state loses contact with the strong language, and then evolves its own. The demands of a new nation require the evolution of a separate language to mark its independence and new sense of national identity. Language is a very important badge of nationhood, and can bind disparate tribes together. Its vital importance can be seen in countries with large minorities who speak their own regional tongue rather than the official language of the state itself.

It's worth repeating that if language unites, it also divides. The country with a single language is always more united than one with strong, regional dialects. The example of Spain is worth mentioning again to prove this fact and to illustrate the importance of language. The Basques in that country speak their own languages, as do the Catalans. Both Catalans and Basques prefer their regional accents to Spanish, the official language of the state. Language is the essential bond of nationhood, as can be demonstrated in Spain. The Catalans have been trying to secede and form their own state, although Catalonia has been allowed to govern its affairs. Instead of Spain being united under a single and unifying language, it is divided by different and divisive languages. The official language of the country has not been dominant enough to eradicate strong regional dialects, as the laws of nature demand. There is no alpha male language in Spain to unite the country, and therein lies its intractable problem.

When empires collapse, new states emerge to fill the vacuum because territory is too precious to be left vacant. It happened in Europe when the Roman Empire collapsed, and it happened again when the Soviet Empire collapsed. It also happened in India when the British Empire withdrew from that sub-continent and the country gained its independence. The bitter religious rivalry between Hindus and Muslims that the empire had papered over fractured and tore the country apart in bloody pogroms and acts of brutal violence. Pakistan emerged as a new state with a new national identity and flag. It also required a language that differed from India as a badge of national identity, and adopted Urdu as its official tongue.

India too required a national language to proclaim its independence from the colonizer, and chose Hindi as the official tongue of the new state. It retained English in its courts of law, and also for trade because it's the international language of commerce. Then Pakistan split apart and a new nation emerged calling itself Bangladesh. It adopted Bengali as the official language as a badge of its independence and nationhood. Pakistan and India are now bitter enemies and both are armed with nuclear weapons. India and Pakistan evolved different languages because they were in competition with each other.

The fracture of India into three separate and distinct countries gives compelling credence to the theory that language has evolved under the laws of competition. Nature does not pit different species against each other, but members of the same species under the fourth law. Humans compete with fellow humans for territory and living space, and always have because they are highly territorial. Differences in language identify a tribe or a nation and spurs that competition. It happened in Europe when the Roman Empire fell and it happened in

India when the British withdrew. The nations that emerged from the post-imperial chaos required a national identity, and there is no better badge than language.

These examples are offered as proof that language evolves into separate tongues to represent a new state. Each emergent state was now in competition with rival states, and separated by language. It was an essential badge of nationhood, just as a new flag declared the independence of that nation.

The rise of the nation states in Europe required the shedding of blood to unite different kingdoms and statelets into a single nation under a single ruler. That unity also required a single language as imposed by the strongest ruler. That strongest ruler then imposed his dominant tongue to unify the country. That language acted as a bonding agent to bind the nation together, and enforce its laws across the population. That language also helped heal regional disputes between neighbours who had previously spoken in different dialects. Each nation state was defined by its own badge of identity, by its own language that marked it out as different to its neighbour.

To further indicate how language works, it shall be necessary to make a comparison between the USA and the EU. The USA was united under a single language, that foremost and essential tool for nationhood. The native languages that it encountered were rendered extinct by the stronger competitor as decreed by the natural laws, or forced into tiny pockets with few speakers. The many dialects of the regional tribes were overwhelmed when they met the stronger language. English did more to unite that vast continent than the railway lines that tied the east coast to the west coast, or indeed the flag.

To illustrate how a single language is essential to the unity of a country, a comparison is now made between the Canada and USA. Canada has no single and unifying language to unite the country. There are two official languages, each struggling for the alpha male position. English and French are locked in combat, with neither tongue able to gain the top spot, the alpha male position that is essential for unification. As there is no single and national language, only one outcome was possible and that is clear to see. Canada is not united under a single language, but divided by two and torn between English and French, with no foreseeable victor. In the last century, the French speaking province of Quebec wanted to secede from Canada and go its own independent way, and there is still a strong movement for a break with the rest of the country and declare independence. With two languages seeking to dominate, Canada is not as united as its southern neighbour that is governed by one language.

Now the USA is beginning to face a similar crisis of identity with immigration from South American countries that mainly speak Spanish. Although the children of immigrants are taught English in American schools, many of the people who flow into the country are illegals and keep below the radar. They do not speak the unifying tongue because of their status. There are areas in the state of Florida where Spanish is the dominant tongue, and there are signs in both Spanish and English across the southern United States with large Latin populations. Any society with two competing languages is prone to fracture and split into divided communities. Without the unifying bond of a common tongue, the glue of society weakens and the unity of a nation begins to tear apart.

One of the reasons for the foundation of the European Union was to prevent another war on the continent. Ostensibly, it began as a trading organization, but the ultimate goal has always been to create a unified continent to rival the USA. In other words, European countries would become like states in the USA, united under a common flag and national anthem. The principle of a united Europe is a noble ideal, even if the political classes have managed to conceal this endgame from the people. The EU has established some building blocks of this latest attempt to unify Europe, this time by peaceful means in contrast to the conquerors who tried it by force of arms. It has a common flag, an essential requirement for any union that diverse populations can live under. It also has a common anthem, possibly the best one around, Beethoven's Ode to Joy. However, it lacks the cornerstone that supports the edifice, and that is a common language. It can never emulate the success and unity of the USA without that supporting cornerstone.

The EU can never be as united as the USA because there are too many languages competing for the alpha male spot, and each strong nation pushes its own language for that coveted position. The countries that form the EU are not united by a single language but divided by many tongues. The less powerful nations, such as Sweden or the Netherlands, do not push their own language because they are not competing for that top spot. Britain is withdrawing from the EU, but the French would never permit English to be the dominant language because that would cede the top spot to a rival, even if it is the international language of trade and travel.

Spanish is a global language, and they would not permit French to get that top spot because that tongue is a rival and competitor. Germany is the most powerful country in the EU and as such its language should represent the continent. However, that subject is taboo because of its history and is not mentioned in the upper echelons of power. As the matter stands, there is no common tongue and therefore no unifying language to bind the citizens together. There is too much national pride at stake to adopt a common language from one of the contenders because it would prove a blow to national pride for the losers. Here is the dilemma facing the political classes and one that they are powerless to solve.

Under the fourth natural law of competition, no single language can be allowed to gain superiority over another. That scenario would suggest that the language is dominant over every other one on the continent. Language is an essential badge of nationhood and status, more important than the national flag and more important than the national anthem. Flags and anthems can be changed with no great fuss, but language is the essential badge of identity that cannot be changed. The largest and most powerful languages on the continent are in competition with each other as nature demands, and one must not be allowed to gain the upper hand. An advantage must be countered under the fourth natural law. If the nations of the EU are ever to be united, a common language is the most essential ingredient to make it happen, just as flour is the essential ingredient in a loaf of bread. Without flour, bread cannot be made. Without a single language, Europe can never be united.

The only time Europe achieved almost total unity was under a single language. When Latin ruled the continent, Europeans were united more than ever before or since. Latin abolished wars and established a peace that lasted for two centuries. The continent had not known peace for that duration before, and would never again in its history. When kingdoms and countries evolved from the ashes of the fallen empire, each developed its own language. Latin was the glue that bound different tribes and peoples together, and when that bonding glue dissolved, they went back to fighting each other for territory. Here is proof that the same language unites, but that different languages result in war.

It is no coincidence therefore that both world wars began in Europe because that continent is divided along the tectonic plates of language. When tectonic plates rub against each other, the result is an earthquake. When two languages rub against each other, the result is conquest and war until one is dominant. No king or emperor has succeeded in imposing a common language on the continent since Latin ruled, and no organization like the EU can ever achieve that goal.

The practical applications of a common language cannot be stated often enough. It would facilitate the movement of trade by eliminating the translation of documents from one European language to another. A common language would save billions of euro paid to translators by the taxpayer to translate the laws and directives that pour out of Brussels like a river in flood. It would facilitate travel and tourism because everyone in the union could understand everyone else, from Latvia to Spain and from Germany to Malta. And yet, children across the EU are not taught a single tongue that would enable them to communicate with fellow Europeans. This is a language hurdle that the politicians are unable to jump over because of entrenched nationalism that they are unwilling or afraid to tackle head on.

When the American colonies fought a war against Britain and the USA was formed, the new country adopted the language of its colonizer. The English language had united the country from east to west, and proved a better bonding agent than anything else. The language had to evolve under the fourth law of competition, and it did. There are some subtle differences in the language, both in spelling and different words for the same item.

Americans spell the word *colour* as *color,* and the word *centre* as *center*. They also refer to a *tap* as a *faucet*. The *wing* of a car is called a *fender*, and the *boot* is a *trunk*. Very subtle and minor differences perhaps, but they do go to illustrate that language evolves to portray separate identities. It depicts the new nation by being different, and as an essential badge of its new identity. These facts would tend to back up the theory that language changes when new nations or states emerge to lead independent lives. There is no other rational conclusion. The case has been proven beyond a reasonable doubt.

SPORT AS THE ALTERNATIVE

Mother Nature decreed that species compete against the same species for valuable resources and territory. In the case of mankind, this competition has led to constant and unremitting war: but it has also led to sport where rivals can compete without inflicting too much injury on their opponents. Sport can best be described as war fought by peaceful means. No other species on the planet competes against each other at sport, but then no other species is as competitive as man.

The forging of the nation state into a single country under a common flag and language ended local wars between groups and tribes, but it did not end the competition between them, nor could not have ended it. The fourth law of nature demands competition between the same species and that command is not open to debate or refusal. The internal wars between tribes and clans evolved and became competitive sports. The bitter rivalries between clans and tribes over territory were now directed into bitter rivalries in sport, and derby matches are a reminder of the hostile origins of sport.

Derby matches between neighbouring counties or parishes are always the most fiercely fought, and are a throwback to the tribal warfare fought over adjoining territories. Land disputes in the past occurred along boundary lines between neighbours, as they still are between hostile nations such as Pakistan and India. Derby matches are fiercely contested because of their origins in border wars between neighbouring tribes. Losing a match against a neighbour is far worse than losing it against a distant rival. The loss is almost as hard to take as the loss of territory. The home team is expected to play its boots off to defend the honour of the home crowd, just as their forbears took up arms to defend their home from a rival tribe. Here is the reason why derby matches are so bitterly contested, because they have their origins in the armed conflicts for land waged by neighbours in the past.

Football in its various forms the most popular sport in the world and the one played by most people across the world, principally because it is a team game. The basic unit of the troop is a team, and football reflects that same ethos. Unlike golf that is an individual appeal only and is elitist, football has a wide appeal that allows the people to participate, and to play for their community or nation. It has particular popularity amongst the young and is played by that age group, the same age group that once fought in the territorial wars between tribes. They now compete against their neighbours and other nations on the playing field, and represent the new warriors who go forth to do battle for their tribe, but without too much bloodshed.

The pride of the tribe is at stake, and beating the opposition gives the members bragging rights whenever they meet off the playing field. The winning team that returns home with a trophy is given a rapturous reception by the people, in the same way that returning armies weighted down with booty were greeted after a victorious campaign abroad. The close links between war and sport can be seen in this simple comparison.

One ball game in particular has been badly tarnished by hooliganism and condemned by critics, without the reason for the violence being questioned or understood. All facets of human nature in the present can be found in man's origins and species and soccer violence is no different. Soccer hooligans are doing what their forebears did, except the game is used as a flag of convenience under which they fight. The hooligans are generally in the same age group as the players, and are fighting for their tribe on the streets and terraces just as their team is competing on the field of play. When they follow the national team abroad and start trouble, they are fighting for their country.

Not all fans fight, but some are more prone to violent conflict when abroad than others. Fans from countries with no tradition of aggression outside national territory are less likely to cause trouble abroad. Fans from countries with a history of aggression or imperialism are more likely to cause trouble when they attend matches abroad. The conflicts that their countries waged whilst at war now evolved into street battles between rival fans in peacetime. The competitive drive between rivals simply found a new channel to express their warlike aggression, using sport as a substitute.

As previously stated, the Olympic Games were used as a device to give breathing space to warring states to recover, before resuming what they did best: fighting each other. That brief period of peace transferred the rivalry from the battleground to the stadium, but it did not curb the need for aggression. Contestants were killed at some of these games and cheating was rife. The Olympic ideal has never fully come to terms with human nature, and mankind's inexhaustible capacity for violence.

The modern Olympic Games also promote an ideal competition between athletes, despite the long history of cheating and drug-taking by athletes and governments alike. The promotors of the games insist that winning is not the goal, but the taking part. A good slogan, except it ignores the competitive urge that is inherent in nature, and it ignores the truth. If winning is not the goal, why is a gold medal awarded to the victor? Why is a silver medal awarded for the second place? Silver is not as precious as gold. Why is a bronze medal awarded for third place? That metal is of lesser value than gold and silver. It would appear that the organizers do not heed their own idealistic slogan, and the games are all about winning. After all, nobody remembers the losers.

Sport reflects life, and just as there are winners there must also be runners-up and losers. If winning is not the ultimate goal, there would be no need to cheat nor take drugs. Competitive sport can create friendship between nations, but it can also lead to real conflict. In 1969, Honduras launched a military attack against El Salvador over a qualifying match for the World Cup. Competing against other nations has become a source of national pride for the winner, reflecting its importance.

Winning a World Cup can boost a nation's own self-esteem and its sense of importance on the world stage. Its leaders can bask in the reflected glory of victory, and the cup creates a sense of pride in the nation as a whole. The modern Olympics are now used for these purposes, and have more to do with a country's image abroad than promoting love and friendship between nations. That is why some countries have a state policy of cheating, and deny it if found out. The games are not about taking part, but winning at all costs as their history reveals.

Every Olympics produce a league table of winners and losers, and that in itself reflects how the world works. The most powerful countries seek that coveted alpha male position at the top of the table and regard it as a goal of national pride. China announced its arrival on the world stage by topping the medal league it staged in Beijing. Its goal now is to emulate that feat where it matters most, by becoming the alpha male military nation. The close linkage between sport and war can also be observed in the staging of the Olympic Games in Berlin, where Germany declared its arrival as a world power. The games proved to be a trial run for Hitler's ambition to become the top military power on the planet. Germany dominated the Berlin Games, and three years later set out to conquer the world, this time with tanks and planes.

Just as the original Olympic Games offered only a brief respite between warring states, the modern ones are no different. Wars will always be fought between nations because all wars are for territory, and always shall be until the end of time, or the end of the world. That is the second natural law and it has dominated the mind of man since he first evolved, and it remains as powerful as ever in the psyche of mankind. The rivalry between the Greek city states was not ended by the Olympic Games, but merely postponed the endless wars to another day. The rivalry ceased only when Rome conquered Greece and imposed a strong alpha male rule that the warring states could not impose.

This is a refrain that has echoes in other European countries. German rival states fought wars against each other before that country was unified under a strong leader, and Italian city states fought wars against each other before it was unified. The wars ceased when the countries were unified under a single leader, giving the people a sense of national identity, but now external enemies had to be fought. Sport replaced war within national boundaries between rival statelets, but not between rival nations. The 1936 Olympics did not prevent World War Two, but rather set it in progress. The Berlin games was the powder keg that set the whole world ablaze.

Having achieved top position in the games, Hitler set out to gain top position in the world, having proved Nazi superiority in the stadium, the same young athletes who had ran in races now flew planes and drove tanks. Hitler declared that his nation required more territory, or as he put it more living space for the German people. It was one of the few times in his life that he actually told the truth. The war against Russia was fought principally for more territory, in the same way as every war since the dawn of mankind has been fought for more land.

Single combat in war has evolved into peaceful competition in sport, although some contact sports can be dangerous. A boxer can get killed in the ring, but that hasn't stopped the sport because it attracts a worldwide audience and remains a subliminal reminder of mankind's rise to dominance. Men fight because they have evolved for conflict, and boxing channels that natural aggression into the ring where there are rules. The widespread interest in boxing has lasted for centuries, and it was part of the first Olympic Games. Heavyweights grab the most attention and they command the biggest purse of any division because he is regarded as the ultimate alpha male who can defeat any other man on the planet. As a result, the heavyweight title is respected above all others and holds a fascination for the world. The reason lies in man's evolution at that period when brawn ruled, and the best fighter could gain the alpha male leadership of the tribe by trials of physical strength.

In the same way that kings fought kings on the field of battle, primitive troop leaders fought their counterparts from rival troops. The winner then took control of the territory and often the rival's troop. The winner had proved his fitness to rule by trial of combat, and his better genes would make the next generation stronger. The heavyweight championship of the world is a title worth fighting for, and the history of the sport proves winner can have his pick of ladies. They attend the fights and cheer on their favourite. Fighting was the way that primitive leaders gained access to the females, having to prove their fitness in this manner.

Fighting for access to females ceased only when mankind changed from brawn to brain, but the legacy of that era still remains. The message is latent and subliminal, but it is there nonetheless. The past never really goes away, but is retained in the memory of mankind.

Sport evolved as an attempt to replace war. It did not succeed because territory was more important, and that meant it had to be fought for by strength of arms. Territorial disputes could not have been resolved on the field of play, but on the battle grounds of war in armed conflict. Territory was literally a matter of life and death and both sides realised that they had to decide the issue by killing members of the rival tribe, or dying themselves. Sport does not require that same sacrifice because there is only pride at stake. Sport forms close friendships between individual athletes, but individuals are not nations. They compete against each other for medals and for glory, and for national pride. But competing for territory is another matter entirely and requires a more violent and aggressive approach. Territory is much more important than all the glory and all the medals in the world, and that message is hardwired in the human brain. That is the reason why war has always trumped sport and always shall.

Losing at sport is a temporary loss of face, but losing at war can end in death.

THE CYBER FUTURE

As clearly demonstrated in this work and proven beyond any conceivable evidence to the contrary, every war since the dawn of time has been about territory regardless of the alleged motives. Man is the most territorial animal ever to have stalked the planet, and by far the most aggressive and violent. The territorial imperatives of land ownership, under the second law of nature, have caused him to establish visual markers of his own patch. Whether his territory is defined by a river, by a mountain range, or by an artificial dyke or wall, that barrier is a visual warning against trespass. In other words, the territory is already occupied by another tribe who shall defend it to the death. That is a territorial law first established by Mother Nature and one that all humans fully understand. The law is part of their condition as primates. It need not be taught in school, since humans are born with the law indelibly imprinted on their minds as territorial creatures.

No activity undertaken by humans evolves faster than war, but now the rulebook has been torn up. Man has evolved so quickly that he has superseded the long-established rules of nature and moved beyond them. Just as his move into farming distanced him from the rest of the animal kingdom, his latest move is taking him deeply into the unknown abyss. Now he has moved further away from his roots by moving into the realm of cyber space. The old and certain laws of territorial ownership are being eroded and diffused as he enters cyber space in his constant search to gain an advantage over his rivals. The warning signs of rivers and mountain ranges are becoming obsolete because the war of ideas is now being fought on the airways and in cyber space. That is the new frontier, and it's ideal because there are no boundaries or warning signs to deter him. There is no visible deterrent in cyber space to warn off an aggressor, like a river or a national flag. The territorial laws of nature have fallen to the warlike tendencies of mankind, and there is no going back.

War began on the ground with battles between tribes for more territory, with the stronger tribe overrunning the weaker tribe and seizing its territory. The quest for more land is hardwired into the human brain and cannot be altered nor removed. That quest gave rise to empires and the violation of frontiers that sparked wars. Then it moved to the sea when man constructed warships to fight his enemies and acquire more territory. Ships allowed him to seize territory abroad and claim it as his own. Then war moved to the air when he constructed planes to fight his enemies.

As if three theatres of warfare were not enough, now man has moved into cyber space to wage war against his enemies. Cyber space is the new battleground, the fourth theatre of war, and the most dangerous thus far in his evolution because there are no boundaries. And, of course, there shall always be enemies out there to fight under the law of competition between the same species. It cannot be otherwise because Mother Nature has laid down the rules for life on Earth.

Cyber space is perhaps the best theatre in which to fight because it provides many advantages. Man can use his most powerful organ to fight it, that same organ which granted him mastery of the planet, that ingenious organ between his ears. The cyber war is currently being fought with thoughts, and there is no more destructive weapon on the planet. The competing ideologies are in competition for the minds of men, with no missiles or bullets being exchanged, as yet. Opposing ideologies are fighting it out using propaganda, mistruths, conspiracy theories, and black ops to gain an advantage over the enemy. State players are involved under the umbrella of private citizens that provides cover for their operations, and gives the state plausible deniability. If the players are exposed, the state can deny responsibility and protest they are private individuals with access to a computer. This fourth theatre of warfare does not require regimented soldiers in uniforms and boots; it can be fought by geeks dressed in runners and shorts.

The battleground is not between truth and lies or right versus wrong, but between the most compelling and persuasive ideologies. Truth has always been in the ear of the beholder, and it remains what the ideology claims it to be. Truth is not a universal concept based on a solid and immovable foundation, but a shifting perception that the listener believes to be true. No truth is absolute in the war of propaganda and spin. Right and wrong are subjective to the stronger ideology. The indoctrinated are unable to differentiate between these two opposites.

Cyber space is the new frontier, and mankind has readily moved into arena of warfare as easily as he has fought on land on sea. There are no frontiers in cyber space, no flags to mark out national territory, nor no river to tell an enemy that he is entering a land owned by another tribe. An enemy can enter that space without being checked or halted, nor indeed questioned. He has free movement to come when he pleases, and stay for as long as he desires because the authorities do not know he has transgressed the national frontiers. The breach of airspace by an enemy plane is spotted by radar and jets sent aloft to warn it off, and shoot it down if it ignores the warning. Under the same universal rules of engagement, an enemy vessel that enters the territorial waters of another country is warned to retreat, and is likely to be fired upon if it ignores the warning.

There are no similar boundaries in cyber space, and the enemy can recruit members of the tribe who live there and turn them against their own leaders. That enemy can sow the seeds of discord from cyber space without fear of retaliation, simply because no rules have been established to replace the old rules of engagement. Entering cyberspace is not an act of war because the intruder has breached no visible warning signs that are universally recognized and fully understood. The old and established order of territorial markings do not apply in cyber space. This is the new Wild West where every state player can behave as he wishes because there is no sheriff to enforce the law. To put it more bluntly, there is no law to enforce since no crime has been committed.

An attack by a nuclear device can be countered by a retaliatory strike, but how can an attack from the airwaves be countered? That is the new theatre of war, and it is already occupied. An enemy can plant propaganda from cyber space and claim it is the truth and nothing but the truth. It can reinforce the belief that many people in the West hold that all politicians are crooks and liars. Cyber space propaganda can ram home the message held by many that Western politicians are only in politics to line their own pockets, and don't give a damn about the people who elected them. There is a current phrase that states truth is alternative facts. Since there cannot be two opposing truths or facts, the brain is forced to choose between them. The indoctrinated mind shall always, always, always choose the version that is implanted in its memory banks and that reinforces what the person already thinks. The message that confirms what a man or woman believes to be truth shall receive a happy welcome, and shall not be open to question.

Cyber space is also changing the nature of spying. Now there is no need to meet discreetly in parks or hotels to recruit from the opposing side, spies can win over on the airwaves. There is no need for cash to change hands or bribes because a strong ideology is the most effective recruiting sergeant in history, and the cheapest. There is no need to hand over wads of cash for information because the indoctrinated person will hand it over at no cost, and not have a twinge of conscience because he or she has bought into the ideology. The person's memory banks have been corrupted, and he or she is blissfully unaware of the infection.

Ideology transcends class and colour, people of no educations and college graduates. It transcends rich and poor alike. The nuclear secrets and launch codes of the enemy can be obtained if spies are recruited in the military. The strong ideology is more potent than all the nuclear weapons in the world because it can render them obsolete overnight.

The goal of using the enemy's nuclear weapons against the indoctrinated person's own cities is a distinct possibility. If this possible scenario sounds like science fiction, it must not be forgotten that science fiction sometimes becomes science fact. If such an action were to happen, the deterrent value of nuclear weapons is lost because there is no external enemy to strike back at, nor to launch a retaliatory strike against. The military top brass cannot know who is responsible for the attack, nor what country has recruited the spies except that their cities have been wiped out. They have been preparing for an external strike, not for an internal launch against their own cities. No military is ever prepared for the unexpected attack from a new source because it invariably takes lessons from the last conflict. The weaponry that has cost countless billions can be rendered useless overnight by the use of ideology. It has lost its credibility, but more importantly it has lost its irreplaceable deterrent currency.

What effectively the move into cyber warfare has achieved is to render all that has gone before obsolete. The rulebook has been incinerated by mankind as nations continue to compete against each other for more and more territory in a world of exploding population growth and a finite acreage of land. The greater the numbers of humans on the planet, the greater demand for territory. That fact is inescapable, except for the usual suspects who prefer platitudes to truth. As a result, the problem of exploding human numbers is seldom addressed because the subject is considered taboo. Better to sweep the problem under the carpet than engage in a rational conversation.

The territorial animals that man shares the planet with still abide by the second natural law as they have since they first evolved. Lions and wolves still use scent to mark out territory and warn off potential rivals. Chimpanzees in the wild still use visual displays at the boundaries of their territory, aided by auditory warnings by stomping their feet against tree trunks. The nation states that now exist behind clearly defined warning signs and barriers and national flags are no longer secure because the enemy has bypassed these visual warning signs by moving into cyber space. Everything has changed utterly, and a terrible new world is coming into existence. Just as the invention of the cannon rendered castles obsolete, the invention of cyber warfare has rendered boundary markings and national flags equally obsolete. There are no auditory nor visual warning signs in cyber space to warn off an enemy and tell him he has intruded upon a territory owned by someone else. The enemy can enter when he pleases without hearing a firing shot or a word of warning.

From a state of certainty where every global player knew the rules, the world has now become a place of uncertainty where no player knows the rules. This limbo existence can be illustrated by two incidents, the first that happened in the last century and the second that occurred in this century. When, in 1941, Japan attacked the USA at Pearl Harbour the response was immediate. The USA declared war on Japan for invading its territory and striking at its naval facilities. These were the rules that nations understood in that era, an attack would be met with a similar or more powerful response. Contrast the action taken by the USA in the last century to its response to an attack in this century.

When, in 2016, Russia attacked the American elections, there was no response. It used cyber warfare to interfere with the elections and helped get the candidate of its choice elected. There was no similar attack mounted against Russia in response. It had managed decide the outcome of the election in its favour without firing a shot. An armed attack by Russia would have brought an immediate response, but the cyber-attack, went unanswered because there were no rules for this fourth theatre of war. And there are still no rules for this most dangerous theatre.

Cyber warfare represents a total and complete rupture with the long-established laws of Mother Nature that have been existence before man evolved on the planet. Now there are no rules because no visual warning signs exist to warn off intruders. For the most territorial animal that has ever inhabited the planet, this lack of clarity and certainty is an intolerable situation, and one he is struggling to adequately deal with. Now no country knows where it stands. Now an enemy can invade its space because there are no deterrents to warn the aggressor off. As highly territorial animals, primates are programmed to defend their turf, but how can they defend it against an attack from cyber space? The old certainties are gone and have not been replaced. The old order has changed utterly, and left an intolerable and dangerous vacuum in its wake.

A mechanism or device shall have to be invented to proclaim ownership of cyber space, just as national symbols mark out terrestrial ownership. Cyber space must be clarified because it involves national security, and there is no more pressing concern than defence of home turf. Otherwise, a time shall surely come when the threat posed by cyber warfare shall result in hard conflict with multiple human deaths. And since man tore up the rulebook of Mother Nature, it is incumbent that he must solve the problem he has created.

Once more, the consequences of man's interference with the rules of Mother Nature can clearly be seen.

THE LOSS OF SPIRITUALITY

In the first decades of the third millennium CE, mankind stands unchallenged as master of his world in every sphere. That fact is indisputable. However, with power comes extra responsibility; but man's dominant position as master of every species on the planet has rendered him domineering. In his hubris, he has come to believe that he has a right to exploit other life forms for his own benefit instead of sharing it with his fellow creatures. He continues to deplete the planet's natural resources and gives nothing back in return.

The respect for nature by our earliest ancestors has been lost by modern man on his way to the top. It needs to be addressed before more life forms on the planet are rendered extinct by modern man's insatiable greed and total disregard for other species. There is a custom still practised in Ireland by old men who harvest turf by hand to use as winter fuel. They offer the first sod to the bog by throwing it over their shoulder in appreciation for the gift. That courtesy and respect for a gift of nature has been lost as the planet is denuded of its forests, its rivers and lakes polluted, and its resources depleted to the point of no return. A person who receives a gift without offering thanks is regarded as disrespectful and crass;

but man does it daily as he plunders the natural world, and is universally lauded when he gets rich.

The oil industry is but one example of an industry driven by greed and maximizing profits. Oil spills and pollution are considered risks worth taking to maximize profits, instead of ensuring that the extraction poses no threat to the lives of creatures who share the planet, nor the irreparable damage to the environment. The mantra that 'greed is good' dominates the thinking of oil companies, just as man believes he deserves to dominate the world. The oil company that strips resources for profit alone and has no care for the environment must be condemned as looters by civilized society. No resource supplied by Mother Nature should be taken without offering thanks for her generous gift.

The growth of super cities is increasingly the way mankind has chosen to live, but in so doing has further distanced itself from the natural world. Cities offer better opportunities for people to find employment and rear their families; but man does not live by bread alone. He is, above all else, a spiritual being who also requires bread for his soul, and that is called spirituality. He needs that for his wellbeing and for a healthy state of mind. No city can grant the serene sense of spirituality that a woodland or a forest can endow.

City dwellers are not living a natural environment but in a manufactured one that is far removed from their ancestors. No amount of concrete and glass can ever replace a tree that is a living organism, and no city can substitute for a forest where peace of mind can be found. Lifeless canyons of tarmac and steel and concrete do not bear comparison to a spread of trees in full bloom and filled with life. The traffic of the city grates on the ears, but the songs of birds in the forest are uplifting and bring a feeling of contentment that no vehicle can replace. And no amount of material wealth can quench that thirst for spirituality. Man has lost that vital essence of his being as he distanced himself from nature.

Modern offices in a city contain computers and chairs, and many also have plants and shrubs. These are now regarded as part of the furniture, but they are much more important because they are a direct connection with man's origins. They are a subliminal reminder of a former life spent in a forest. The past never really goes away because it reminds man of his origins. Plants also relieve the stresses of modern life, and many people who live in city apartments have window boxes to retain that link with nature. That connection is still there because it remains part of the journey to the modern world.

Every society that has ever existed anywhere on the planet has sought spirituality, which demonstrates that man requires this divine feeling to attain happiness. Every society that has ever existed invented gods to explain how the world worked, but as importantly to channel that spirituality and commune man with his driving force. Man has created the idea of a soul to highlight the difference between the quest for bodily welfare and the search for spiritual fulfilment. He also assumes that he is the only life form on the planet who has one, and that he is unique and special. The soul is not part of the body, although it exists in some obscure place and form within the body. The soul can be likened to the lifeforce that animates the person, and answers the question about the meaning of life. That is a question still being debated widely by philosophers, and has been since man learned to think. Perhaps the meaning of life can be interpreted as the spiritual quest to find happiness.

Pilgrimages to holy places are still happening, even in the modern world that has become largely secular. Man has created 'holy cities' where he believes his god visits, and who has favoured them over every other city. These 'holy cities' are thronged with religious tourists and pilgrims in search of spirituality. Usually, these cities are places of conflict because they do not cater to humanity as a whole, but to a particular religious group. Where different religions lay claim to the same city, such as Jerusalem, bloodshed is certain to happen because each side believes the city is their exclusive holy place and that the competition has no right to claim as its own. To put it more clearly, it's more about territory and power than spirituality and enlightenment. Spirituality should be inclusive to all human beings, not exclusive as in organized religions.

Now many people are discovering the first religion that regarded Earth as the mother goddess. The Earth religions of the past are making a comeback as man continues to seek bread for his soul. Sites like Stonehenge and Newgrange act as magnets on winter and summer solstices, and draw large crowds who commune with nature and the changing seasons. The worship of Mother Nature has never really gone away, and lies deep within every human being because she is the original mother who gave them birth. That reverence is part of the human condition, dormant but still there, and reawakened in many on the days of the solstices. There is a spirituality in these ancient sites that's totally lacking in organized religions, and in their places of worship.

The reverence for Mother Nature continues to exist in societies cut off from the modern world. The respect for nature has always been part of hunter-gatherer societies because they understood how the world worked. They lived in harmony with their environment, taking only what was necessary to sustain life. Being closely connected with nature provided sustenance for their sense of wellbeing that modern man has lost on his way to the top. That sense of serene spirituality is not something money can buy, it's priceless. It can be found in nowhere else except in the green spaces of the natural world.

The reason for that the loss of spirituality began with the discovery of farming, and that discovery distanced man from nature and his roots. He became civilized, but in the process had lost his aura of spirituality and that vital and essential spiritual connection with the natural world and its understanding. He began to regard the place of his initial birth as an alien and hostile environment that must be tamed, and went about his task by slash-and-burn tactics to clear the forests. The animals that he deemed a threat to his crops and livestock were vilified, before being hunted down and killed. Nature was not a friend who supplied all his needs, but a wild and savage beast that must be tamed.

Another contributing factor in the lack of respect for the natural world can be found in monotheism, in the religions that preach and believe that their single god created Earth and all its life forms. As a direct result, these religions decreed the creator god of all things gave his followers domination over every other creature. Judaism and its two offspring Christianity and Islam are anti-nature religions in the extreme, and deem any worship of nature as paganism. The worship of the natural world is contrary to monotheism, and these three religions have vilified the worship of nature, and hunted down and killed pagans whenever or wherever they found them.

The terror that monotheism inflicted on paganism has lasted for over two millennia, and yet is seldom discussed or brought up in forums where genocide is discussed. The greatest civilizations of the past were pagan, and the greatest thinkers. The classical civilizations of Egypt and Rome and Greece were pagan. The people who deified the wonders of nature by worshipping them as gods and goddesses were vilified because they had a different belief system. There could be only a single ruling dogma when monotheism took control, and anyone who did not accept the new order could expect little mercy. The vilification of paganism has reverberated down the present day, and it still bears a stigma in the minds of many followers of these religions.

Eastern religions are different because they are more tolerant, and consequently lack the fanaticism that is inherent to monotheism. Hinduism, Buddhism, and Jainism are basically nature religions, and as a result wildlife is treated with more respect in in the societies where these religions are practised. Followers of Jainism share their crops with migrating storks and snow geese flying south to escape the harsh Siberian cold. Buddhists monks feed wild animals in their temples, an act that would be seen as blasphemous in the three monotheistic religions. The cities of India, where Hinduism is the principal religion, teem with wildlife that co-exist with humans. It goes largely unnoticed by the people because they are closer to nature, but the sight of a fox in a Western city can make the newspapers and TV channels.

The concept that man was given dominion over all aspects of the natural world was at total odds with pagan religions that treated nature with due reverence and respect. In Celtic religion, rivers and lakes were inhabited by goddesses and offered tribute in the form of precious objects. And people do not abuse a resource they deem sacred, but instead treat it with respect and care. The pollution of rivers and streams by modern man is a direct result of that lack of respect and disregard for nature. The oak tree is another example of the Celtic respect for the natural world. The tree as treated as a sacred object and worshipped by the priestly class, not as a resource to be plundered but for religious ceremonies. It performed an essential role in their religion that allowed them to commune with their gods.

Greek religion too worshipped nature by creating gods and goddesses to represent the wonders of the natural world. A sacred grove dedicated to a goddess would have been cared for by the worshippers of the deity, not cut down nor burned and desecrated. This reverence and due respect disappeared when monotheism waged an unremitting and savage war on paganism simply because its followers worshipped nature, and drove it to near extinction. The ancient religion is only now making a comeback. If paganism can regain the foothold in the minds of people it once enjoyed, then a respect for the natural world can be rekindled.

The sundering of man's roots uprooted him from his forest origins and led him to the city. He lost contact with Mother Nature who had brought forth his species, and broke his umbilical cord to the past and his origins. The spirituality of the forest was mislaid when he came to live in the city. Modern man has become a lost and wandering soul without a star to guide him, tossed on a stormy sea and seeking a port where he can find safety.

That port is called nature, except he has now become so filled with hubris that he believes modernity and science can fulfil his spiritual needs. These are merely his comfort blankets that are incapable of warming his soul, or granting him the lost spirituality he seeks most to give him peace of mind. Neither science nor any manmade religion can satisfy his needs unless they recognize nature and regard it with the reverence they give to their gods. A reconnection with his roots in nature represents his only way forward, and the true road to happiness. He must seek the embrace of Mother Nature, just as his forbears did many millennia ago.

Modern man has disturbed the fragile and delicate balance of nature due to his ignorance of how the world works, and the way all forms of life are interconnected and interdependent. The destruction of a single tree affects the myriad of creatures who depend on it for food and living accommodation. A tree removes carbon, a greenhouse gas, from the atmosphere and replaces it with oxygen, thereby preventing global warming. The loss of a single tree also disturbs the food chain by removing a habitat for insects and birds. Birds play an essential role in the balance of life by keeping insects under control, and by distributing the seeds of the tree in their guano, which acts as a fertilizer to help grow the next generation.

Forests are disappearing quickly due to commercial interests and the need to grow more food. Recently, in one year alone eight thousand square kilometres were cleared from the Amazon rainforest to grow crops and raise cattle. Forests in Asia are being cleared to grow palm oil, and threatening the viable future of many life forms, including orangutans who are primate relatives of mankind. Global warming was the predictable result of man's total disregard for the natural world and his interference with its delicate balance. Beneficial insects like bees are disappearing at an alarming rate due to rapid climate change, and modern man is incapable of understanding that the disappearance of these insects represents a grave danger to his own existence. Insects pollinate the fruit orchards and the crops he needs to sustain life. It does not take a genius to foretell the future if the destruction of wild habitats continues at the present rate.

As this work has demonstrated, when man interferes with the natural balance, he causes havoc. Nature is a finely tuned machine with each part working in harmony. When man interferes with one part of the delicate machine, it stops working. The law of unintended consequences then kicks in and causes unforeseen problems. It happened when he hunted the wolf to near extinction, and it happened when he used poisonous chemicals to control lifeforms he regarded as pests. Poisoning a crow does not only kill the bird, it also poisons the food chain and kills any animal that feeds on the crow.

The most pressing problem the world now faces is global warming. As usual, not all nations are in agreement because each nation is in competition with others. Whether or not man can reverse global warming is another matter, but it hardly seems likely. That requires cooperation between nations, but man is the most competitive animal to have ever existed on Earth. Already there are politicians who contradict scientists that sound warning signals about the planet heating up at an alarming rate. Politicians do not think in the long term, but rather in the present and staying in power.

There are also politicians who welcome the melting icecaps in the Arctic Ocean because it opens up new sea routes to accommodate international trade. Their motive is to control the sea lanes and grab more territory to claim as their own. Territory has always been the best game in town, after all. The competition between the usual suspects has now moved to that part of the planet for its rich resources, and for its strategic possibilities. That attitude is akin to a group of poker players on the Titanic ocean liner arguing over the winning hand whilst the ship sinks below the sea. Life on Earth is under threat, but the chance of riches and strategic advantage over potential enemies on the world stage trumps everything else. The planet and its life forms are in danger from climate change, and yet the prospect of territorial gains is deemed of more importance than melting icecaps.

As the world becomes warmer, weather conditions are certain to become more extreme. Floods will be more frequent causing people to lose homes and livelihoods, and that process is happening across the world. The consequences of flooding will inevitably lead to relocations of communities, and compel them to find safer ground, causing friction with settled communities. The melting of the polar icecaps will bring similar problems to coastal cities, since water always follows the path of least resistance. In the same manner that war drives people from their homes and turns them into refugees, global warming is likely to create climate refugees, but on a much larger scale. Conflicts might be avoided if the refugees are part of the country where the cities are submerged, but if they seek refuge in countries already suffering, they will not be welcomed. The barriers will go up because land is a finite resource and not to be shared with outsiders.

If the above scenario appears alarmist, it should be pointed out that climate refugees have already abandoned their homes and sought refuge elsewhere. Over the past two decades, five of the Solomon Islands have sank below the waters of the Pacific Ocean due to global warming. They were part of an archipelago of low-lying islands that are threatened by rising sea levels. The islands are sparsely inhabited and therefore unlikely to cause mass refugees; but the flooding of a major coastal city would create a flood of people seeking safety. The pun is quite intentional. As with Europe and as with America, refugees are tolerated in small numbers. They are seen as a threat when the numbers look like an invasion force.

The United Nations has estimated that a million species face extinction due to the threat caused by a warming planet, not to mention the disruption caused to other animals who have to seek a colder environment. Fish such as cod can survive only in colder waters. Animals can usually adapt to a changing climate, but they need time because change comes gradually. The planet is warming so quickly that they may not have time to adapt, but become extinct. Mankind directly rendered other creatures extinct like the dodo and the carrier pigeon; but now he indirectly is threatening the survival of a million species.

The statistics make for sober reading, except for people who regard climate change as a hoax, and nations who seek to benefit from melting polar caps and rising sea levels. It has been estimated that forty per cent of amphibians will be in danger of extinction if the planet continues to warm up. Amphibians have been on Earth far longer than mankind, and humans owe a duty of care to protect them.

There can be no way of forecasting the result of their extinction, except to say that all life on Earth is inter-connected in some way that is not yet fully understood. Life on Earth can be compared to the web of a spider, a tug on one strand sends vibrations across the web. Fourteen percent of birdlife is on the danger list, or fourteen out of one hundred different species. Is that the type of legacy that the current generation wishes to hand down to children yet unborn?

Possibly the greatest threat the world faces lies in the oceans, the original birthplace of life. Sharks and rays are estimated to decline by as much thirty-one per cent. The knock-on effects to the essential health of the oceans will be devastating if these numbers come to pass. Sharks are at the top of the food chain, and in that role perform a vital service to keep the oceans clean and healthy, plus the species they prey upon. They remove the sick and diseased; and also perform the role of undertakers by disposing of dead carcasses. Without them, the oceans would quickly become a cesspool of diseased fish and rotting carcasses left to fester and pollute the waters.

One third of the world's coral reefs is under threat from global warming. The reefs can be compared to cities beneath the sea where fish can find food and safety, and give birth to the next generation. The reefs offer protection for smaller fish where they can hide from predators, and they can breed in relative safety. The death of coral reefs can be compared to withdrawing oxygen from a city of humans. Without oxygen, the people die and the city dies. Without coral reefs, the fish die and the next generation is lost. The symbiotic relationship between reef and fish is essential to both because they rely mutually on each other for existence.

Twenty-five per cent of mammals are on the danger list, and humans are also mammals. Tigers are in a perilous state already due to loss of habit due to increasing human populations, and also the superstitious belief that their bones are a cure for many ailments. It's difficult to imagine or conceive a world without these magnificent animals but it could happen. The prospect of future generations viewing images of tigers as we view dinosaurs is a real possibility.

Future generations shall not forgive the present one if tigers join the long list of species that man has driven to extinction by greed and disrespect for all life that he shares the planet with. Man must rediscover that awe and harmony with nature that his forbears had. Most importantly of all, modern man must reignite the spirituality and respect for the natural world that has been extinguished in his rise to the top.

The destruction of the planet and the complex web of life shall continue unless humans reconnect with Mother Nature, and learn to respect the planet that gave them life. Humankind must not view Earth as a resource for endless plunder and looting, but as a home that should be passed down to the next generation in good order, and a healthy place to live. No responsible parent would set fire to the house where his child was raised, nor leave it in a worse state than he found it. Instead, the parent would ensure the house was in good working order, free from structural faults and in the best condition for those who come after him. The wise parent knows that the house is not his own, but belongs to the next generation, and to the one after that. He is simply the caretaker who protects the home during his lifetime.

A worldwide and intensive tree-planting programme should be instituted to halt global warming, with each country given a target of trees to plant. This could be achieved according to the population of a country so that all countries could participate. For instance, a country of five million people could be encouraged to plant ten trees per head of population over a specific period until reaching an agreed target, with incentives added for every fifty million extra trees planted.

Each country therefore would have a shared and collective responsibility to combat the problem of global warming, thereby instilling a shared and global sense of the importance to protect the planet. The talking shop known as the United Nations should declare an emergency to underline the urgent threat to the planet and its life forms. Poorer countries could be given aid to plant trees in areas with soil erosion, which would also stabilize the soil and create more land for farming. Planting a tree and watching it grow would help people bond with the natural world, and rediscover the awe humanity once held for the planet and for Mother Nature. The respect for the natural world that the first humans had must be recovered, not only for the future of Earth but also for the spirituality of mankind.

The clock is ticking.

THE PRACTICE OF EUGENICS

Evolution is a curve of continuous improvement whereby successful genes are honed and improved, and passed down to the next generation. The unsuccessful genes are removed or allowed to die out, thereby ensuring that the next generation is healthier. The laboratory of life that is part of the evolutionary process sometimes produces accidents of birth, such as malformed offspring or life forms that have little chance of surviving to sexual maturity. Every new birth is both a trial and an experiment under the process of *neotento*, and sometimes this process produces offspring with physical or mental defects. These are removed quickly to prevent their genes being propagated. Mother Nature has evolved the predator which plays an important role in ensuring that only the best genes are passed on to the next generation. That is the specific role she has decreed for the predator, to weed out and remove genes that should not be propagated.

The calf born to a wildebeest mother on the African plains that is weak or deformed is removed by predators, ensuring it does not reach maturity. The next generation must not inherit defective genes, and that is a golden rule of propagation. The herd itself is tested by running from predators such as wild dogs who pick off the old and weak stragglers that are unable to stick the pace. By these continuous trials of endurance, the weak are exposed and removed, and the fit are allowed to breed. The health and vitality of the next generation is improved as a result, giving the offspring a better chance of survival and ensuring a healthier population.

The perpetuation of good genes and the removal of bad genes is the eternal dance between predator and prey that is understood by both actors in the drama because they are part of the natural process of eugenics. Springbok antelopes leap and prance to show the predator that they have good genes and therefore should not be targeted. The weak are exposed to attack by their inability to leap as high or prance as much as their fitter rivals. The fitter animals are not targeted by the predator since they are seen to have good genes, and as a result the predator concentrates on the weaker members of the herd. Predators exist in the air, in the oceans, on savannahs, in jungles, and in every other sphere where life exists. Their roles are specific, they are the surgeons of Mother Nature evolved by her to perform a specific role. They remove the weak and diseased, as a human surgeon cuts out a malignant growth from a patient. If a human surgeon does not remove a malignant growth, the patient dies. If predators do not remove defective genes, the animal populations they prey on will also die. Mother Nature practises and promotes eugenics, and predators are her surgeons who keep flora and fauna healthy.

Man too, went through this phase of his initial evolution, and although he has become civilized and sophisticated, he too is still part of the natural world. He carries inside his head a subliminal message from his past about eugenics, and that only the best genes should be passed down to the next generation. Modern man has distanced himself from Mother Nature to such an extent on the road to civilization that he does not realise the message still governs how he thinks and how he behaves. Some examples are worth noting here to demonstrate that he is not in control of his actions.

In the ancient city state of Sparta, babies were examined at birth for defects by the elders of the community. Any babies found with them were exposed on a mountain and left to die, allowing their weak genes to die with them rather than perpetuating them. Sparta was a militaristic society based on the strength of its fighting men, and weakness was not tolerated. Only those deemed the fittest and strongest were allowed to survive into adulthood and sire the next healthy generation. The subliminal message that only the best genes should be passed down dictated who should live and who should die. The rulers of Sparta were hearing a call from the primeval past that only the best genes should survive, although they were unaware of the message.

The Nazis actively practised eugenics by murdering people with mental and physical defects in their bid to create a master race. The purpose was not only to create a perfect race without blemish, but more importantly to remove them from the gene pool by preventing them from breeding. Their perceived weak genes were a threat to the idea of a master race that had to be without physical and mental blemish. This was the underlying and subliminal reason that influenced their thinking, the removal of defective genes to prevent them being passed on to the next generation. The Nazis too were hearing that subliminal message from distant past when man lived in close harmony with nature. They were not the only ones to target the less fortunate in society. Other countries were also influenced by the third law that states only successful genes should be propagated, and that bad genes should be extinguished.

The USA sterilised its citizens who were deemed mentally unfit to breed, although these actions have been conveniently swept under the carpet of memory. Russia too engaged in this practice, as did both Japan and Sweden. The Nazis have been rightly condemned for their evil deeds, but there is a collective loss of memory from countries who deem themselves more civilized, and do not wish to be associated in any way with the vile practices of Nazi Germany. There is a continuing and persistent pattern here that cannot be ignored nor disregarded as mere coincidence. The subliminal message of eugenics as promoted by Mother

Nature is imprinted on the mind of man, and always has been because eugenics is part of the natural world.

Not only have the mentally ill been removed from the gene pool to prevent them breeding, the message of eugenics has also affected those with a physical disability. The pursuit of perfection has long been part of human culture because it remains part of nature since its roots are firmly planted there. It was not too long ago that people were labelled *cripples* or *hunchbacks* and other pejorative names because of their physical defects. These terms are now rightly regarded as an obscenity against the person; but the thoughts behind the insults cannot be erased because they form part of man's evolutionary journey from the past to the present. Genes are the driving force of life, the engine that propels it forward, and the propagation of the best genes decides who should breed and who should not breed.

The lame person is not responsible for his lameness of course, which is what the ignorant insults suggest, nor the deformed woman for her deformity. These are accidents of birth that the laboratory of life produces now and then under the process of *neotento,* which is continuous trial and error with each new birth. Under this process, each new birth is an experiment, not a repetition. That no two life forms are identical proves this important point. Yet the belief that a man with a spinal deformity had somehow earned it persists even in our more sophisticated world. The widespread belief that he was somehow responsible for the deformity is ridiculous and ludicrous, but it persists because the subliminal message of eugenics is part of human nature. It remains part of human nature because man is part of the natural world. The deformed man is not regarded as good husband material because of his perceived bad genes. He might be pitied by the opposite sex, but rejected as a mate. The genetic imperative that drives life demands that only the best genes must be passed on, and the children of a deformed man might be born with the same condition.

That internal message cannot be ignored nor disregarded because it is part of nature. The third law requires humans to pass on successful genes, and the person without a physical defect is regarded as having better genes than the person with a physical defect, and therefore can produce good offspring. Mates are not chosen randomly, but decided by which partner can give the next generation the best start in life. The philosophers who argue about the meaning of life should not engage in abstruse thought, but examine how nature works. Genes are the meaning of life because they are the driving force of life. The next generation must be better than the present generation by the perpetuation of the best genes.

The third natural law governs all life on the planet, and determines who should be allowed to breed and who should not. The handsome man is more attractive to women than the ugly man because the former is regarded as having better genes. The beautiful woman is more attractive to the opposite sex because she is seen as the bearer of good genes. The perpetuation of good genes is the primary reason behind the attraction of the sexes. Men and women do not view potential mates with their eyes but with their genes, and who offers the best prospect of their continuation. The message that good looks are a sure indicator of good genes, and shall be passed down to the next generation is one that cannot be ignored because it is part of the natural world. Again, some examples are worth mentioning to prove this point.

The male lion with a large mane and a good coat is a more attractive choice to females than the lion with a small mane and a sickly coat. The former is seen as the bearer of healthy genes whereas the latter is regarded as the bearer of unhealthy genes. In a similar manner, male birds of paradise compete with each other in the beauty stakes to attract a mate. The male that can grow the most colourful display of feathers is regarded as having the best genes. The females choose their mates because they rear the next generation, and they choose the best mate to give their young the best start in life.

The third law is universal and applies to both fauna and flora, and proves that all life obeys the third law whether a flower or a bird. Flowers compete with each other to attract pollinators by putting on bright displays of colour and advertising by scent. The flower with the best display or the most compelling scent attracts the pollinators, and has its genes perpetuated. The sickly flower that is unable to compete does not attract the pollinators, and its genes are extinguished. The flowers therefore must compete with each other to attract the pollinators by adopting new colours and scents, or have their genes extinguished. This competition spurs evolution and acts to improve the next floral generation.

The third law that propels the natural world also drives man because he is part of that world, even if he thinks he has free will. The message that good genes are determined by beauty and bad genes were determined by ugliness influences how he thinks and how he behaves. It matters not that beautiful people can produce ugly offspring and ugly people can produce beautiful offspring; the subliminal message is so powerful that it cannot be ignored nor contradicted. This same message is common to all peoples and all societies, demonstrating that the laws of Mother Nature are universal to all mankind. Beauty is globally prized and praised, and its opposite is condemned and avoided.

Studies have shown that where two persons of equal ability apply for a job, the more handsome or beautiful person is far more likely to get that job. The link between beauty and goodness is as old as recorded time and continues into the present. For instance, gods are always handsome because they are good, and therefore carry the best genes. Devils and demons are always ugly because they are bad, and therefore carry the worst genes. People embrace gods because they are handsome and reject devils because they are ugly, and that is a common theme down the ages in most societies. This is a constant message that repeats in art and literature, and reinforces the subliminal message of good genes and bad genes. This same message resonates in art too down the ages.

The roof of the Sistine Chapel in Rome, painted by Michelangelo, portrays a terrifying scene from the Last Judgement where the wicked are condemned to Hell. The devils are grotesque and with contorted features, portraying them as evil. They are evil because they are portrayed in that ugly light. The angels are good because they are handsome, and therefore the bearers of good genes. God is benign because he is handsome, and Heaven is filled with the good and the beautiful. Hell is the opposite, and a place where the ugly dwell because they have bad genes. The Devil is an ugly beast with bad genes, and is often portrayed with horns and cloven hooves. He is seen to be evil precisely because he is ugly, and therefore the bearer of bad genes.

The message that beauty equals goodness and ugliness equals evil has entered the world of fairy tales, and children too are influenced by eugenics since they too are part of the natural world. Whether in old fairy tales or modern movies, monsters are evil and wicked because they are invariably most fearsome to behold. In the fairy tales that frighten children, evil witches must be ugly and good witches must be lovely, the essential and universal requirement for distinguishing good from evil. The prince who slays the dragon or the monster must be handsome to behold, and his victory is a triumph of beauty over ugliness.

To put it in evolutionary terms, the victory of beauty over ugliness represents the triumph of good genes over bad genes. The human mind finds it impossible to accept that monsters can be handsome because it contravenes the subliminal message of eugenics, just as it finds it difficult to accept that princes can be ugly.

It should be noted that the most evil men in history, were not deformed or ugly, but men who looked much the same as other men. Hitler and Stalin were no pin-ups, but they could not be called ugly either. Evil resides not in the body of man, but in the mind of man. That is the true home of evil and wickedness, but the message persists because it remains part of the natural world. The idea that bad genes manifest themselves in ugliness or deformity is so powerful that it cannot be erased from the human psyche. It remains embedded in the human mind because of the importance of the perpetuation of genes.

When a mass murderer or serial killer is caught, the usual comment is that he looked much the same as any man on the street. People expect him to stand out by being ugly or deformed, such is the power of eugenics. That the public at large is attracted to beauty and repulsed by ugliness is a matter of fact and passes without much comment. The beauty industry is founded on the perception that a handsome face is a passport that can open all doors and allow its wearer to live happily ever after. The winners of Miss World are praised for their beauty, as if they had earned it like a law degree or a PhD from a prestigious college boasting huge fees. That they have not earned their beauty but inherited it from a combination of genes is not mentioned and goes without notice.

Handsome film actors too appear on the pages of magazines, and are used to sell everything from alcohol to zithers. Beauty sells because beauty attracts. Ugly does not sell product because ugliness repels. No agency worth its salt would hire an ugly person to sell a product because of the genetic message of bad genes. Man has become civilized but he is still part of the natural world, although he believes he can think rationally. He is still influenced by eugenics and by the message that beauty equals good genes and ugliness equals bad genes. That perception remains because it is part of nature.

Mankind has always abused and discriminated against those people in their community with a physical or mental defect, and the reason is the subliminal message of eugenics. The mentally ill were chained up in madhouses and left to die so that their genes would be extinguished and not passed down to the next generation. Even in this more modern and supposedly enlightened era, there are regular reports in the news of the mentally ill being abused by their carers who have a duty of care to look after them. There is an explanation for all aspects of human nature and it can usually be found in nature. Yet modern man has become so detached from his origins that he never seeks answers in the right place, and comes up with his own theories to explain why the mentally ill are abused by people who should be ensuring that their days and nights should be peaceful. Man looks to the stars as his final journey, but he does not recall his origins or where he came from, nor want to remember.

The people with physical defects suffered as badly and were driven from towns and villages to fend for themselves away from the world of 'normal' people. The freak shows of the nineteenth century saw them transported like circus animals to amuse the gaping crowds, and were often pelted with rotten fruit. Even in death they were not spared, and they were exhibited like stuffed specimens for the amusement of the crowds. The underlying reason for this inhumane and terrible treatment was the third law that only the best genes must be passed on. Physical deformity was regarded as proof that these unfortunates were the bearers of bad genes and therefore not worthy of propagation. By treating them as outcasts and aliens, they were removed from the breeding population.

Children display signs of their origins as they prepare for life as adults. The reason they play is a legacy from an era when play-fighting established a pecking order, and built the cohesion of the troop. The modern child with a physical deformity can expect to be bullied or abused at school, or shunned by classmates. It happens with a regularity that cannot be otherwise explained except by their evolutionary past. Children too obey the message of eugenics because it is part of nature, just as they are part of the natural world. Children are not taught to hate and beat up those who are different, but in fact, are taught the opposite by concerned schoolteachers and parents. Yet the subliminal message of eugenics is imprinted on their brain. That is what they listen to, not the good advice from parents or from schoolteachers. The third law rules them just as it does all other forms of life on the planet.

The human beauty industry is also founded on the need to attract a mate, and is not consigned to the female sex alone. Both sexes supplement what Mother Nature has given them with the use of artificial aids such as skin creams, hair dyes, and makeup. Looking the best is also part of nature to attract a mate, and more extreme forms such as plastic surgery are also used to enhance beauty. Enhanced lips, buttocks, breasts, and noses can make the person more attractive to the opposite sex. Plastic surgeons are now integral to the entertainment business. The need to attract is part of nature, and therefore part of mankind, and is responsible for an industry that's worth billions in international currencies across the industry.

An essential part of that industry is perfumes. Women wear perfume and men use aftershave with the same underlying goal in mind, to attract a mate just as the flower uses it to attract a pollinator. Scent has been relegated by mankind, but it still plays an important role in the mating game because it forms part of mankind's origins. Pleasant smells attracted in the past and unpleasant smells repulsed, just as they do today. The lesser importance of smell has not altered the part the nose plays in attracting the opposite sex. It can still differentiate between good and bad odours, and be either attracted by the scent it picks up or repelled by that scent. The success or failure of an expensive perfume depends on the scent that is most pleasing to the sense of smell.

It can reasonably be assumed that the need to attract a mate began when man's ancestors abandoned the troop or tribe and started living in pairs. There would have been no need to attract under the harem system when the alpha male had access to every female and lesser ranked males were excluded from the breeding stakes because the dominant male controlled the females in his harem. Teaming up in pairs required keen competition for mates, and that endeavour required the art of attraction. Women were now in competition for the best man to team up with, and men too were in competition for a suitable woman to give his children the best start in life. Now any advantage that could attract a mate was employed, like painting the face or wearing flowers to mask unpleasant odours. The competition between males and the competition between females for mates was essential to perpetuate the best genes as demanded by nature.

When leaders adopted royal titles that gave them higher status than commoners, that advantage made their offspring more attractive. Their higher position was a symbol of their better genes in the minds of the people. Even in the modern world, many young girls dream of marrying a handsome prince, the stuff of dreams and fairy tales, and living happily ever after. Princes are handsome in fairy tales because of their good genes, and that is still the case in the real world. The perception that they have better genes because of their high status still persists in the modern era, despite the body of evidence to the contrary.

The belief that princes are good husband material is a legacy from a time when royalty was regarded to have the best genes, and that erroneous belief has not gone away. Royals were also believed to have blue blood in their veins, unlike the common people who had red blood. Blue blood made them more desirable as marriage partners because it was another sign of better genes. Royals do not have better genes than commoners because that erroneous belief runs contrary to the fourth natural law. The primate family tree does not grow royal families, nor dynasties of rule where power is passed on to the next generation. Instead, the primate family produces competition for the top job, and always has.

Royals are a human invention to ensure rule was kept in the family. Titles passed down through the bloodline to the next generation, thus ensuring it had the best start in life. The reason why the reign of kings did not stand the test of time is that it contravened the laws of nature; and as a result the few remaining monarchies are not important on the world stage, and are mostly ceremonial with little or no power. They have passed their sell-by date and the remaining royal families are destined to fall because there is no change at the top. The faces might change but not the genes, and that basic requirement is demanded under the primate system of rule. New blood must be introduced by a changed leader in order to strengthen the gene pool. Princes and princesses might be in high demand as suitable marriage partners, but they belong to an era that has passed forever and shall not return.

    The need to look good is part of the human condition because of the necessity to attract a mate. Clothes were worn by early man to keep warm and stay alive, but they too have evolved to function as an aid to beauty and to show the wearer in the best light. The annual Oscars where awards are handed out to actors have as much to do with fashion as they have with acting, or with the business of cinema and putting bums on seats. The actresses compete with each to wear the most beautiful dress or gown to make them more attractive, just as nature demands. They are, in effect, advertising their good genes for the whole world to see. They too hear the siren call of the third natural law by displaying themselves in the best light. Just as the loveliest flowers attract the most bees, the loveliest dress attracts the most gossip and ensure its wearer grabs the headlines across the world.

The fashion industry itself is not founded on the need to keep warm as clothes were originally intended to do, but to enhance desirability. Good clothes portray the wearer as desirable, and creates that feel-good factor that the person feels when wearing the latest fashion. The compulsion to advertise good genes transcends flora and fauna, and proves that mankind is part of the world of nature. As a result, clothes and beauty products and aids to stop the ageing process shall always be part of humanity. There is an innate need in human beings to attract because the third law governs thoughts and actions.

Mother Nature is still in control of human activities and human thoughts. She has been from the beginning.

MAN INVENTS THE GODS

Was mankind created by an omnipotent deity in the sky? Or did mankind evolve from a simian ancestor in a distant forest in the content we now know as Africa? The choice must be one based on choice or on a personal system of belief. This chapter shall concern itself with the widespread and universal belief in supernatural deities with total powers that transcend the limited abilities of human beings. How did the belief in gods originate?

The quest to discover the origins of mankind is not new and has been ongoing since man first stood on two legs and gazed up at the stars in the night sky with awe and wonderment. That ever-demanding organ between his ears kept asking questions and seeking answers, as it does in modern man. Constant inquiry accelerated the evolution of man's best asset, the brain that had given him mastery of the planet. The question of how mankind came to live on the planet is as relevant today as it always was. The question of how man had come into existence then changed to what he was doing on planet Earth. Eventually, this pursuit would lead him to question the true purpose of his existence and his place in nature. Ancient man speculated on the origins of Earth, and answered the question his brain demanded by coming up with the creation myth. These myths are universal and cross continents, proving that early mankind was as fascinated about his origins as many modern scientists.

Japanese creation myth portrayed a universe in chaos and silent at the beginning. Then a sound was heard that formed material particles. These separated and the lighter particles ascended to become light whilst the heavier particles combined to become solid matter that eventually formed Earth. Chinese creation myth also portrayed a universe in chaos at the beginning from which two elements were conceived known as Yin and Yang. These twin elements represented the male and female, or hard and soft in contrast.

The Greeks came up with the novel idea that the universe created the gods, not the other way around. Ancient Egyptians conceived the idea that the world began in water, which is now widely accepted by science. As the waters receded, the top of a mountain appeared and the sun shone for the first time. Creation myth can therefore be seen as early mankind's attempt to discover the origins of Earth. One particular creation myth would form the foundation stone of three world religions, and it can be found in the Bible.

The Garden of Eden story is accepted as literal fact by countless millions, but it too is a creation myth. The story has all the elements necessary for myth, with an omnipotent supernatural being performing feats of magic. Creation myths should be regarded as early man's attempt to discover his origins, but this one has been widely promoted as literal truth by religious propagandists in search of new recruits. In fact, to even question this myth during the Dark Ages in Europe resulted in burning at the stake. This myth was used to maintain the power of the Church in Europe, just as the ideologies of fascism and communism are used to keep people who still live under them in submission.

Early man invented the gods as a way to explain the inexplicable world of natural phenomena that he could not otherwise understand. He questioned what sort of being was hurling bolts of lightning from above that destroyed his home and livestock. He did not know that lightning is caused by electrical discharges in the atmosphere that often trigger a bolt. He did not know that lightning strikes randomly and the strike that destroyed his home was not a personal attack against his life. His brain demanded an answer for the terrible phenomenon that threatened his life, to enable him take precautions and stay safe. The first law of survival demanded that he must take measures to escape the lightning bolts. Who controlled that awesome power and why was it directed at his home? As his brain progressed, it kept demanding more information to answer such questions, in much the same way that a muscle *demands* more exercise to improve and get stronger. That brain was his best asset and had to meet all challenges that came his way in order to survive. Lightning or a sudden earth tremor had to be understood by his brain to be countered. His existence required him to find answers to dangers he could not otherwise explain, such as those presented by a predator.

Since lightning bolts or earth tremors were beyond his capacity to understand, he put them down to beings with supernatural powers. These beings, or gods, did not live by mortal rules that he could understand, but by their own rules, by supernatural rules. These beings could visit death on anyone who offended them by hurling bolts of fire from the sky, or shaking the ground beneath their feet. Early man's lack of knowledge could have steered him in no other direction. He did not know, as modern man knows, how lightning is caused. He did not know, as modern man knows, that ground tremors are caused by earthquakes as the planet's tectonic plates slip and collide against each other. His village or settlement could be swallowed down when the ground beneath his feet opened up and he stared into the abyss. For that reason, his brain demanded answers to the threat posed by this phenomenon that posed a threat to his life and that of his family. A look at the Greek myths can give us an insight to how he solved these problems.

Zeus was the chief Greek god, the alpha male, and his family mirrored the society that had invented it. He had a wife and brothers and sisters. Zeus used lightning or his thunder bolt to punish any mere mortal who offended him. Therefore, he had to be appeased with rich offerings at his temple to prevent him hurling his thunder bolt at a shepherd tending his flocks on a hillside, or at a farmer sowing his grain. Gold was offered to placate the god, and church collections at churches on Sundays are a legacy of this appeasement. His brother Poseidon resided in the ocean, and it was asking for trouble to undertake a voyage without appeasing this capricious god before setting sail. He could stir up the ocean to a frenzy and capsize the ship, killing everyone on board. How could any ship captain trade goods across the ocean with this vindictive god lying in wait? He did what his modern counterpart does when trading goods by sea, he took out an insurance policy to protect his livelihood.

When a massive volcanic eruption obliterated the island of Santorini about three thousand years ago, a giant tsunami wave devastated the island of Crete and wiped out the Minoan civilization. Here was Europe's first advanced civilization, with double-storied houses, ordered streets, and apartments. Its art was truly remarkable and lifelike, far superior to static Egyptian art, and featuring bright murals of dancing girls and bull-leaping men adorning the walls. The ancient artists who painted the murals captured the movements of the dancers and gymnasts that might have been painted during the late Impressionist movement on the same continent.

Recently, Cretan archaeologists have unearthed pottery shards thrown up by the tsunami depicting images of a trident. The trident was an image associated with only one deity, and that was Poseidon, the god prone to acts of violence. The shards are remnants of vessels that were once filled with gifts to appease this most mercurial god of the sea. Somehow, the people had offended Poseidon and sought to buy him off by offering gifts to assuage his terrible anger. Here was the insurance policy that the seafarers paid to Poseidon to escape his wrath.

Appeasement is a recurring theme that runs through most religions, from paganism to monotheism. If the gods are not appeased, they will rain havoc down on mortals. A worse fate can follow too if mortals do not respect them or honour them constantly with gifts. The gods can withdraw their protection and not offer the people aid and comfort in their hour of need. They can turn a deaf ear to their pleas and withdraw their protection, and leave the people helpless and open to attack. Under these circumstances, an insurance policy is worth taking out by appeasing the gods with gifts and prayers.

Everything that is living evolves including thought, and therefore religion changed as the thoughts of its followers changed to meet the evolving lifestyles of its followers. All ancient religions were pagan, and this word is derived from the Latin *paganus* which means villager or rustic. Paganism was essentially the worship of the forces of nature, or a deep respect for Mother Nature that had survived since man first invented the gods. He invented crop gods to ensure a good harvest, and home gods to ensure it was a safe place for his family. There were river goddesses and tree spirits that reflected the harmony he felt with nature and his surroundings. There were ceremonies to mark the changing of the seasons, and they were personified by gods and goddesses. And where goddesses are worshipped, there also is nature worshipped because that too is feminine.

The evolution in thought eventually led to monotheism, or the belief in a single deity. To be more specific, monotheism evolved with a male deity as the sole god, and would mark another rift with Mother Nature and a respect for the natural world. Monotheism swept away all the old pagan gods and particularly the goddesses. Monotheism is a masculine religion in contrast to paganism which is a feminine religion. The powers invested in goddesses under paganism were usurped by the single male deity who ruled alone and without a wife in the sky. The change would create an unbridgeable chasm with nature because the male had replaced the female aspect of nature. In future, the goddesses that had personified the forces of Mother Nature would not be worshipped, and the pagans who continued to worship them were hunted down and killed.

The three branches of monotheism that evolved from paganism gave man dominion over every other animal, not as their protector but as their master. The command that first appeared in Judaism was quite clear, that man had sole authority over every creature that the deity had created. This command entered its two offspring, Christianity and Islam. It should be pointed out that authority does not equate with respect, and these religions do not respect the natural world. The awe and reverence that paganism had for fellow creatures was lost forever when monotheism took firm control of the minds of men. Monotheism was, and remains, anti-nature both in outlook and in practice.

Monotheism did not spring fully formed from the human mind, but evolved from earlier forms of worship already in existence. The evolution of monotheism, with the belief in a single male deity of unlimited powers, first occurred in Egypt under Pharaoh Akhenaten about three and a half thousand years ago. This ruler is now known as the heretic pharaoh because he replaced the many gods of Egypt with a single god known as the Aten. The god was portrayed as a sun disk who shone his rays directly on the pharaoh and his family.

The reason why the pharaoh decided to abolish the many gods of Egypt and replace them with a single deity has not been discovered. The most plausible explanation is that the priests were becoming too rich and too powerful. They were probably challenging the king's sole authority too, and no ruler can let that happen and hope to retain power. The priestly class has always enriched itself as conduits to the gods. Their principal claim to power is that they can intercede with the gods on the behalf of supplicants, and pave the way for the life of happiness in the hereafter. That also gives them enormous influence and power, and no ruler likes to have his authority questioned. There can be but one alpha male in a kingdom, and a threat to that sole authority must be countered. That is how the primate system works and has always worked on every continent, whether religious rule or secular rule.

Monotheism failed in Egypt, but it would take root in Israel and form the basis of three major world religions. Eventually, this movement would evolve to become the theocracy, or rule of the priests. Monotheism has always been more concerned with acquiring and retaining power in this world than preparing the faithful for the spiritual journey to the next world. Although Judaism, Christianity, and Islam vehemently claim to be strictly monotheistic, they too evolved from ancient pagan beliefs, and it can be proven that all three have roots in that system of nature worship.

Everything that is living evolves, and that is especially true of thought. That too has evolved as man's brain developed and he acquired more information. Thought does not spring fully formed into the mind but builds on thoughts that are already there, just as Christian churches were built on pagan temples. The foundation stone of monotheism is predicated on a male deity who rules alone. If he is married, the foundation stone is built on quicksand that cannot bear the weight of the temple it supports. Only pagan deities get married, not a bachelor god who rules alone and without a consort. Yet there is strong evidence to suggest the Jewish god had a wife. Her name was Asherah and she is mentioned in the Book of Jeremiah as the Queen of Heaven. This title was directly borrowed from pagan beliefs and entered the Jewish religion. Hera was the wife of Zeus and she was known as the Queen of Heaven. Isis, the Egyptian goddess, also held this title. In Roman mythology, Juno was married to Jupiter the chief god, and she also held this title. Hera, Isis, and Juno were pagan queen goddesses married to the chief pagan deity. Here is but one example of many to prove monotheism has pagan origins.

Christianity too borrowed this title from paganism and attributed the title to Mary, mother of Jesus. Islam also borrowed heavily from paganism, and the name of its god predates the evolution of the religion. Allah was a chief god who ruled over three hundred and sixty idols, one approximately for every day of the year. His symbol was a crescent moon because the Arabs were lunar worshippers. With the evolution of the religion, the idols were dispensed with, but the crescent moon was retained. That has now become the international logo for the faith.

Thought evolves because it lives, but the changes are gradual. There are few if any gigantic leaps in the train of evolutionary thought, but a single link added each time. This progress can be likened to the journey of many travellers from the past to the present. One traveller completes the early part of the journey before handing over the baton of knowledge to the next one.

Each traveller on the journey adds to the combined knowledge as the relay continues forward. The sum of knowledge is gradually built up as the relay continues and the baton handed over. Neurons work in much the same way as each new piece of information is added to the knowledge already stored in the brain, which in turn helps to advance the thought processes.

In this manner, the more acceptable elements of pagan religions were retained and some were dropped. That is how the festivals of Christmas and Easter entered the Christian religion, whilst the blood sacrifice of animals was dropped. There was no gigantic leap from paganism to monotheism, nor could there be. Monotheism was constructed on the thoughts already laid down, and they were pagan. Here was evolution at work as paganism evolved and changed to monotheism. These subtle changes made the transition by the followers easier to accept. Their belief system was not jolted by a sudden rupture with paganism, but evolved over time in line with their thoughts.

A closer look at religions reveals another common thread that runs through them, an essential ingredient in the same way that bread cannot be made without flour. This most essential ingredient pertains to all religions since they were first conceived, whether paganism or monotheism. The chief god must be an alpha male. The irony that religion too must adhere to the primate system of rule is no doubt lost on the followers who believe they are a creation apart. The concept of a female as chief god did not, nor indeed could not, enter the minds of the framers of religion, whether polytheism or monotheism. They could readily believe in the supernatural, but not in the unnatural. And having a female as chief god was far too unnatural to contemplate. The dominant thought would not permit that outrageous concept. The primate system of rule, which demands a male leader at the top, cannot even contemplate a female in this position of power.

Man created the gods as great supernatural beings who were not only omnipotent but omniscient. The gods had all the answers, and they could perform magical deeds that no mortal can hope to achieve. They were created to dwell above common humanity in every way by living in mountain tops or the sky. Here is another clue from animal behaviour that is common to many species, and demonstrates that the gods have their origins in nature. When a male member of a wolf pack approaches the alpha male, he must show subservience by appearing lower. Should he approach without appearing lower, the alpha male regards his approach as a threat to his dominant position and attacks. The strict rules of behaviour are understood by both parties, and the individual who does not obey them is attacked. The leader must always be above the followers.

By creating the gods and placing them in the sky, man was proclaiming them as leaders and acknowledging their right to rule. Primates too obey the visual code of hierarchy by acknowledging the alpha male's status as leader. Should a lower ranked male approach the alpha male in a troop, he must exhibit visual signs of subservient respect. He has to acknowledge the alpha male's right to rule by appearing lower, and he does by making himself smaller. If he approaches tall and erect, the alpha male will regard that as a threat to his dominance and therefore his authority. Such challenges to his authority will not be tolerated, and dealt with accordingly.

Humans too acknowledge their masters in many ways, and placing them on a higher level is one way of showing that deference. A king sits on a high throne that renders him higher than his subjects, or is carried on high to demonstrate his divine right to rule. Humans have also invented appropriate language to show their obeisance to those who rule and therefore entitled to their respect. Although we now mostly live in a democratic age, the legacy of the past is most difficult to cast aside as this work has proven. For instance, a commoner meeting a royal person bows the head or curtsies to affirm the royal's higher authority and status. Making oneself smaller when meeting royalty is a visible sign of the higher status of royalty. Acknowledging a higher authority has entered the language too, although usually unnoticed. A person meeting a king is expected to address him as *Your Highness*. The title alone explains it all.

It is consequently no coincidence that the chief god must live above mankind. He lives on a mountain top where no humans can live, or he lives in the sky because nowhere is higher than that. He must always be above mankind, and shown to be above mankind. If he were shown to live below mankind, he could not possibly rule humans and they could not obey him. Proof of this natural acknowledgement of rule can be observed in how evil gods live. In contrast to good gods, evil gods must always live below mankind. The natural order does not permit humans to worship any god who is lower than them, simply because he is not recognized as their leader.

Evil gods live below mankind to demonstrate they are unfit for rule. This most natural of arrangements can be seen in the relative positions of God and the Devil in monotheism. God is good and worthy of rule because he dwells overhead in the sky. The Devil is bad because he is evil and unfit for rule, and that is why he lives below. Anyone who dares worship the Devil is bad because that is regarded as an act of betrayal against the anointed leader of the people.

Religion requires the suspension of disbelief and as such can be compared to a child's fairy tale. The element of magic that children take for granted can be found in religion, the belief in events that cannot be rationally explained is part of both fairy tales and religion. The critical difference is that fairy tales are seen for what they are as the child progresses into adulthood, but religion is often accepted as truth. The supernatural endows religion with a credibility that it could not otherwise achieve. Magic is as essential to religion as yeast is to bake a loaf of bread. The religion without feats of supernatural magic is not believable, because the gods must be portrayed as superhuman beings with magical powers.

This supernatural legacy that every religion has inherited has come down from a distant past when earliest man invented the gods. He had to explain the natural phenomena that might threaten his survival. Any force or phenomenon that he could not comprehend had to be figured out to enable him mount a defence, as required under the law of self-preservation. Man came up with the idea of gods who performed supernatural feats such as hurling lightning bolts, opening up the ground beneath their feet, creating waves that wiped out cities, and similar feats of magic that they found incomprehensible. Only supernatural beings could perform these mighty works and deeds, and they could also create Earth from nothing except thought. As a result, early man worshipped the gods he had created, in the hope that his life would be spared when bolts of fire were hurled from the sky. He appeased the gods of the sea he had created to give him safe passage on his trading voyages.

Religion remains yet another reminder of man's early origins that journeyed with him from that early primitive age to the present. On that journey, it changed and evolved as the thoughts of its advocates evolved. Whether it has been a force for good or a force for evil is still open to debate. It can be used for good and it can be used for evil. It can be used to ground people into submission, and it can be used to comfort people when at their lowest ebb. Religion is not unlike any other instrument; as good or as evil as its exponents and practitioners. In the final analysis, the decision rests with the individual. As this work has shown, the individual does not have free will. Mother Nature pulls the strings, and has always.

END